Mathematical Ecology

Mathematical Ecology

edited by
Peter Allen (Brussels), Werner Ebeling (Berlin),
Manfred Peschel (Berlin), Peter Schuster (Vienna),
Yuri M. Svirezhev (Moscow)

Mathematical Ecology deals with mathematical models of evolution processes, in the biosphere as a unity of growth and structure-building, with spring-up of new species, their interaction and possible extinction, with the impacts of human activities on the environment and the corresponding consequences for the biosphere.

The whole field seems to be a certain amalgam of suitable system philosophy, in which real world phenomena are considered through an ecological pair of spectacles with the help of system methodology and mathematics.

All publications published herein are of interdisciplinary interest for ecology, biology, economy and technical engineering.

Evolution and Optimization

An Introduction to Solving Complex Problems by Replicator Networks

by Hans-Michael Voigt

with 83 Figures

Akademie-Verlag Berlin 1989

Author
Dr. sc. techn. Hans-Michael Voigt
Zentrum für wissenschaftlichen Gerätebau der Akademie der Wissenschaften der DDR,
Berlin

ISBN 3-05-500617-8

Erschienen im Akademie-Verlag Berlin, Leipziger Str. 3—4, Berlin, DDR-1086

Lizenznummer: 202 · 100/416/89
Printed in the German Democratic Republic
Gesamtherstellung: VEB Druckerei „Thomas Müntzer“, Bad Langensalza, DDR-5820
Lektor: Dipl.-Phys. Gisela Lagowitz
LSV 1085
Bestellnummer: 763 927 4 (9177)
03800

To the memory of my father

Johannes Voigt

who guided my interest to engineering sciences

There is much territory between economics and biology (genetic evolutionary selection) that is still virgin ground. We should not be surprised if the explorations are both crude and pretentious. Wisdom and maturity are the last settlers to arrive in pioneering communities.

Paul A. Samuelson

Preface

The primary goal of this work is to present an introduction to solving complex problems by selection pressure controlled replicator networks.

Doing this we proceed in the following way:

In Section 1. the motivation for this book is presented. Section 2. is devoted to basic properties of the replicator equation. This concerns especially optimization properties. New results on inverse and recurrence replicator equations are included. In Section 3. we consider replication rate controls in detail. Within Section 4. we introduce the diversified replicator equation and show some important properties. In Section 5. we show by means of the graph partitioning problem, the module placement problem, and the travelling salesman problem the application of selection pressure controlled replicator networks. All these problems are *NP*-complete, i.e. hard to solve. These problems were selected because graph partitioning and module placement are of practical relevance, and the travelling salesman problem stands for a prototype of *NP*-complete problems. With Section 6. we give a short introduction to the possible embedding of replicator network algorithms in the design of integrated circuits. Section 7. shows the application of replicator network algorithms to the mapping of process nets to processor networks for multiprocessor computer applications.

In the Appendices two sample programs for the module placement problem and for the travelling salesman problem are included. They are written in BASIC and run with QuickBasic, and TurboBasic as well, on IBM PC's and Compatibles with CGA graphics. You should try them to make your own experiences with replicator network algorithms. Proposals with this respect are highly welcomed.

A vast amount of literature exists in this fields.

For questions on evolution theory the monographs of Hofbauer & Sigmund (1984), who present an excellent treatment of the replicator equation, and of Ebeling & Feistel (1982) are recommended. Peschel & Mende (1983, 1986) developed system-theoretical fundamentals and their applications to systems analysis based on the Lotka-Volterra equation as a unified approach to the modelling of dynamical systems. The book of Ewens (1979) is devoted to the field of mathematical population genetics. The theoretical fundamentals for an understanding of the evolution of macromolecules, i.e. of prebiotic evolution processes, are established by Eigen & Schuster (1979).

As in the case of evolution theory research results in the field of integrated circuit design and parallel computer applications are hard to survey. Here, the monograph of Mead & Conway (1980) and the Proceedings of the Annual Design Automation Conferences and of the European Conferences on Circuit Theory and Design are recommended. Valuable are the Journals VLSI-Design and IEEE Transactions on Computer-Aided Design of Integrated Circuits and Systems as well as Parallel Computing.

The included results were developed when I was with the Forschungsbereich Mathematik und Informatik der Akademie der Wissenschaften der DDR. They were also part of a graduate course on "Fuzzy decision models and structure formation processes" for mathematicians at the Humboldt-Universität at Berlin. The application direction of the results was determined by my former tasks at the Institut für Regelungstechnik.

The content of this book is to a large extent influenced by discussions with colleagues. Here I want to thank especially my academic teacher and colleague Academician Professor M. Peschel from the Akademie der Wissenschaften der DDR who guided my interest to problems of evolution theory. I constantly benefited from the extensive discussions with him. The discussions with Academician Professor W. Ebeling from the Humboldt-Universität at Berlin were always very fruitful. Many colleagues actively participated in the "Volterra-Seminar" at the Forschungsbereich Mathematik und Informatik thus contributing to this book. Especially the cooperation with Dr. W. Jansen and I. Santibanez-Koref from the Akademie der Wissenschaften der DDR is gratefully acknowledged. Discussions with Dr. H. Mühlenbein from the Gesellschaft für Mathematik und Datenverarbeitung about his experiences with replicator network algorithms were very stimulating. Special thanks for the engagement in publishing this book are due to Mrs. G. Lagowitz from the Akademie-Verlag.

Finally, the book could not have been written without the help and assistance of my wife.

Berlin, Summer 1988

Hans-Michael Voigt

Contents

1. Introduction

Questions on the evolution of nature and the human society were always challenging men, and therefore of course sciences.

The 19th century can be marked at all as a century where lasting fundamentals in answering these questions were put down. During this time developed theories had and have impulsive and lasting effects with direct influence on the evolution of human society even during our century.

One of the most outstanding work was done by Charles Darwin and Gregor Mendel. They opened up the door to essential laws of evolution in the biological sciences.

Mathematical modelling in this field was rather rare at this time. The more remarkable is the fact, that Verhulst in 1845 already described a mathematical model for population growth, the well known logistic growth model.

Related to these main streams of biological sciences an enormous progress of research in genetics and ecology was made at the beginning of the 20th century. Therefore it is not astonishing that the twenties and thirties of our century are called the "golden age of theoretical ecology" (Scudo & Ziegler (1978)).

Two names stand for the advancement in this field during this time: Alfred Lotka and Vito Volterra. The famous Lotka-Volterra model for the mathematical description of the dynamic evolution of a population, which consists of a predator population and a prey population, may be considered to be characteristic for this advancement.

Here, an episode for the origination of this model should be presented (Braun (1979)):

Umberto D'Ancona, an Italian biologist, investigated in the twenties the dynamics of the number distribution of different types of fish in the Adriatic Sea. In the course of this research he found out the astonishing phenomenon, that the percentage rate of predator fish measured at the total fishing increased drastically during World War I.

For the Italian harbour Triest the values shown in Figure 1.1 were available for the years 1914 to 1923. D'Ancona raised the question for the influence of fishing intensity which decreased during the years of war supporting thereby the predator fish much more then the prey fish.

The answer to this question contented beside an interesting theoretical meaning for the biologist D'Ancona a very practical one. It included not only an understanding of the struggle for life of competing species but also a guideline for fishery, because fishing methods could not be seen any longer without direct influence on the fishing results.

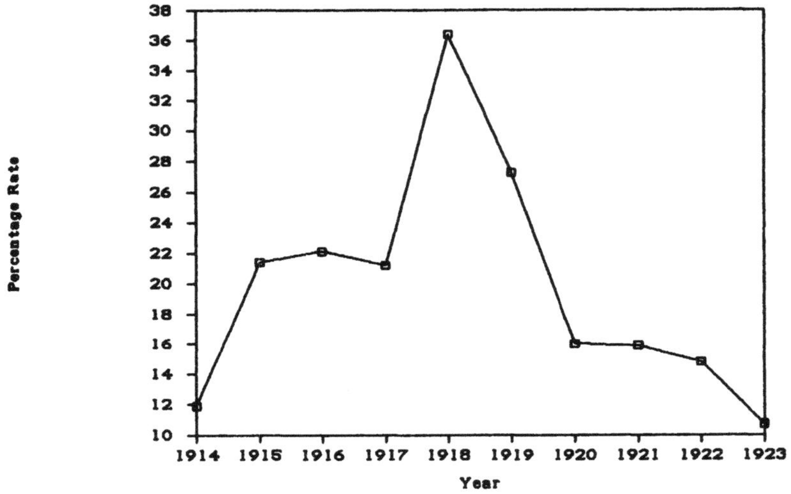

Figure 1.1 Percentage rate of predator fish in the Adriatic Sea

After analyzing the possible biological reasons D'Ancona asked his colleague, the mathematician Vito Volterra, to develop a model of the problem under consideration. The model had the following form

$$\dot{x}=ax-bxy \quad , \quad \dot{y}=-cy+dxy$$

with a, b, c, d = const > 0. Here, x denotes the number of prey fishes, and y the number of predator fishes. The dot denotes the time derivative. By introduction of a fishing rate e for prey fish (ex) and predator fish (ey), one gets as mean values of the prey population, and the predator population, respectively,

$$\bar{x}=\frac{c+e}{d} \quad , \quad \bar{y}=\frac{a-e}{b}$$

It results, that by a suitable fishing ($e < a$) the percentage rate of prey fish can be increased and the percentage rate of predator fish can be decreased. On the contrary, a diminished fishing intensity leads to the support of the predator fish and therefore to a percentage rate increase of them.

This result is known as the Volterra principle. The above given equations are in a somewhat more general form the famous Lotka-Volterra equations.

With this book we want to try to develop a general approach to structure optimization methods which rest heavily on the Lotka-Volterra equations, the fundamental equations of theoretical ecology.

Why? At a first glance such a trial may seem to be misleading!

Let us return to the above considered fishing example. We want to state some working hypothesis and make them operational in form of the following considerations.

(i) The above defined model is characterized by interactions between species (in this case predator population and prey population) which are expressed by interaction coefficients a, b, c, and d. Between species effects of cooperation and competition occur.

(ii) The dynamics of the total population consists of the dynamics of the species. The superposition of interactions will be expressed as growth rates $(dx/dt)/x$ and $(dy/dt)/y$, respectively.

(iii) During a dynamic process stable relations between species take place (as mean values or as isolated fixed points). These relations represent a certain structure of the total population.

(iv) The interaction between the species, and hence the effects of cooperation and competition can be influenced by control variables (in this case the fishing intensity). Dependent on the control values different structures of the total population represented by the stabilized weights of the species may result.

From this point of view a deeper analysis seems to be worth to deduce structure formation methods from theoretical ecology.

Doing this, we will use preferably the replicator equation. This equation will be yielded by a transformation of the Lotka-Volterra equation to the unit simplex.

But we want to go some steps further!

The Lotka-Volterra equation, and the replicator equation, respectively, play the role of a fundamental equation not only in ecology but equally also in population genetics (e.g. Ewens (1979)). The replicator equation serves as a basic model for the assignment of alleles to one gene locus. In this case the characteristic feature is the symmetry and nonnegativity of interactions between allele types. As we will see later, this case is of special interest for the envisioned applications. Hence, the emphasis of population genetics for the unified approach results.

Maynard Smith (1973,1974) analyzed the evolution of conflicts in animal societies. Such evolution is characterized by the genesis of behavioural patterns, and strategies, respectively, of the different animal species. In this case, the underlying model is also given by the replicator equation. The dynamics of the model was introduced by Taylor & Jonker (1978). Within this context , the replicator equation may b interpreted as a dynamic non-zero sum game.

Eigen & Schuster (1979) worked out the theoretical foundatio for prebiotic evolution processes, i.e. for the evolution of macromo ules. The basic model is the replicator equation with a special structu , the well-known hypercycle.

With this considerations in mind, the Lotka-Volterra equatio and the replicator equation, respectively, represent fundamental model for natural selection processes in the sense of Darwin.

To underline this high degree of generality, we use the abstract notion of a "replicator" which was introduced by Dawkins (1982). Following Hofbauer & Sigmund (1984), a replicator is any unit, which may be copied and which shows the following two features:

(i) The nature of this unit must have a certain influence on the probability to be copied.

(ii) The unit must be, at least principally, an ancestor of an indefinitely long series of copies.

Natural selection in the sense of Darwin means survival of the fittest. Simplified, in the language of population genetics this means the increase of the mean total fitness of a population of replicators by mutation and selection. This development to ever higher stages can be understood as an optimization trajectory in an adaptive landscape. Sigmund (1985) showed for the replicator equation a maximum principle using the Shahshahani metric (1979).

At this point we want to raise the question, how to use the present theoretical fundamentals of natural selection for the optimization of "artificial" systems? This question is not a new one. Good results were obtained by Rechenberg (1973) and Schwefel (1977) for parametric optimization problems. Holland (1975) established a framework for genetic algorithms.

But for combinatorial optimization problems (Lawlor (1976), Garey & Johnson (1979), Papadimetriou & Steiglitz (1982)) up to the recent years no working fundamentals were developed which rely on biological evolution mechanisms.

Since approximately 1985 an intensification of research in this field can be observed. So e.g. by Brady (1985) for the travelling salesman problem , the celebrated problem of combinatorial optimization and the theory of *NP*-completeness. Here, also the papers of Goldberg & Lingle (1985) and Grefenstette (1985) on special genetic operators for this problem should be mentioned.

Furthermore it should be noted that since that time good algorithms were developed e.g. for the travelling salesman problem by Mühlenbein, Gorges-Schleuter & Krämer (1988) and for the module placement problem by Cohoon & Paris (1987).

The intensive research on combinatorial optimization and on solving complex problems results from different objective facts:

(i) The graph models which are basic to such optimization problems have in connections with computer applications a wide application range. At present the design of complex integrated circuits, printed circuit boards and finally of computers as well as artificial intelligence applications play a dominating role.

(ii) A large number of combinatorial optimization problems are so-called *NP*-complete problems (Garey & Johnson (1979)). Up to now no polynomial time algorithms could be shown for the exact solution of such problems under the assumption that *P* is not *NP*. The prototype problem of this type is the travelling salesman problem. The computational effort is drastically dependent on the problem size. For n cities there has to be generated and computed $(n-1)!/2$ tours.

Hence, every step in conquering combinatorial optimization problems is of theoretical and practical importance.

This is emphasized by the statistical mechanics approach for the solution of such optimization problems which was developed by Kirkpatrick, Gelatt & Vecchi (1983) and Cerny (1983). The results are impressive, and are nowadays part of almost all good design systems for application specific integrated circuits (Freeman (1985)).

But the conclusions "Simulation of the process of arriving at an optimal design by annealing under control of a schedule is an example of an evolutionary process modelled accurately by purely stochastic means.

In fact, it may be a better model of selection processes in nature then is iterative improvement.", drawn by Kirkpatrick, Gelatt & Vecchi (1983) do not hold in general.

Ebeling & Engel (1986) considered connections and specialities of the thermodynamic approach and the Darwinian approach. They developed a mixed algorithm for the travelling salesman problem which shows good results (Boseniuk, Ebeling & Engel (1987)).

The up to now developed thermodynamic and genetic approaches have a decisive disadvantage. In both cases they are based on Monte-Carlo methods. This leads to the destruction of "good" partial structures by "blind" chance during the stochastic process. All genetic algorithms which avoid this property are problem-dependent, and hence not to be generalized.

In this connection a fuzzy "thermodynamic" approach for partitioning problems by Richardt (1985) should be pointed out. This approach avoids the above mentioned disadvantage.

Within this book we want to develop an iterative approach to the solution of combinatorial optimization problems which is based on biological evolution models, especially on the so-called replicator equation of population genetics. Basic for this approach are the notion of "**replication rate control**" (Voigt (1982)) and of "**diversified replication**" (Voigt (1984)).

2. The Replicator Equation

Living nature produced during the millions of years lasting evolution a fascinating variety of structures reaching from smallest genetic units to large ecological interrelations. And this evolution still proceeds.

For the description of such genetical and ecological relations a vast amount of mathematical models were developed. Dependent on the goals, these models reflect different views and abstraction levels of reality.

Generally, all models are guided by a number of dichotomies which are basic to modelling.

In case of ecological structures some dialectics play a decisive role. Without too much digging into details these are: structure and function; necessity and chance; autonomy and control; cooperation and competition as well as part and whole (Mende, Peschel & Voigt (1983)).

Going back to the beginning of mathematical modelling in the field of genetics, represented by the work of Fisher, Wright, and Haldane, and ecology, represented by the work of Lotka and Volterra, one finds the so-called replicator equation, and the Lotka-Volterra equation, respectively.

These equations are characterized by a very simple structure comprehending at the same time the afore mentioned dialectics.

Not included in the original versions of these equations are the interactions between necessity and chance as well as between autonomy and control.

Considering evolution processes in all its complexity, chance in the sense of mutations has of course a great influence. In the following we want to analyze especially selection processes and coherent structure formation processes. Mutations will be considered in connection with application algorithms.

2.1. The Continuous Replicator Equation

Using the notion of a "replicator" already mentioned in the introduction we may understand by this alleles within a gene pool, animals within a population, strategies of species, and selfreproducing macromolecules. The replication of these units can be modelled by the following general replicator equation

$$\dot{x}_i = x_i[f_i(x_1,\dots,x_n)-\Phi] \quad , \quad i=1,\dots,n \ .$$

It is defined on the unit simplex

$$S_n = \left\{(x_1,\dots,x_n): x_i \geq 0, \sum_i x_i = 1\right\}.$$

In the language of genetics, x_i is the proportion of alleles i within the gene pool, f_i is the mean fitness of alleles i within the gene pool, and

$$\Phi = \sum_i x_i f_i$$

is the mean fitness of the total gene pool. $(dx_i/dt)/x_i$ is the reproduction or replication rate of alleles i.

For further considerations gradient systems (Sigmund (1985)) are of great interest.
If the potential function $V(x)$ is a homogeneous function of degree $s > 0$, i.e.

$$V(tx_1,\dots,tx_n) = t^s V(x_1,\dots,x_n)$$

and the fitness functions f_i are given by the gradients

$$f_i = \frac{\partial V}{\partial x_i}$$

then we have the relation of the mean fitness to the potential function with

$$\Phi = \sum_i x_i f_i = \sum_i x_i \frac{\partial V}{\partial x_i} = sV$$

Linear fitness functions f_i are of special interest because of their analytic properties. In this case one yields

$$f_i(x_1,\ldots,x_n)=\sum_j G_{ij}x_j=(Gx)_i$$

and

$$\Phi=\sum_i x_i\sum_j G_{ij}x_j=\sum_i x_i(Gx)_i=(x,Gx)$$

The G_{ij} are interaction coefficients between alleles i and alleles j.

If we translate "allele" to "strategy" then we get the following game theoretic ansatz:
Let us assume that competing individuals within a population are able to play one out of n strategies. Then x_i is the proportion of individuals with respect to the total population which play strategy i. x_i is normalized to the unit simplex S_n. We denote by int(S_n) the interior of the simplex with $x_i > 0$, and the boundary of the simplex with bnd(S_n).

The game is determined by the score matrix $G = [G_{ij}]$, a real $n.n$-matrix. The scores have to be considered as mean values.

If an individual plays the strategy i against an individual playing strategy j, then the score for the former is G_{ij}. The score for the latter individual is G_{ji}. In general this is a nonzero-sum matrix game. For skew-symmetric matrices we have a zero-sum game.

If the population is sufficiently large the x_i can be considered as probabilities. A consideration as membership values in the sense of fuzzy sets theory is always possible.

Let us assume that the total population is denoted by x. This yields

(i) score for individuals playing strategy i against the total population x is given by

$$\sum_j G_{ij}x_j=(Gx)_i \quad .$$

(ii) score for the total population x against itself is given by

$$\sum_i x_i(Gx)_i=(x,Gx) \quad .$$

If two populations x, y play against each other we have

(iii) score for population x against population y is given by

$$\sum_i x_i(Gy)_i = (x, Gy) \quad .$$

This interpretation of scores implies a number of assumptions:

(i) Every individual plays a fixed pure strategy.

(ii) The offsprings of an individual, playing strategy i, play the same strategy.

(iii) There exists a direct relation between the scores and the reproduction success, i.e. the larger the score is the more offsprings will be yielded.

The introduction of a dynamics into this static consideration rests on the hypothesis, that the reproduction rate for those individuals playing the same strategy is proportional to the score for this strategy. In other words: The reproduction or growth rate of x_i equals the score for strategy i minus the mean score for the total population.

In an analytic form this yields the replicator equation

$$\dot{x}_i = x_i \quad [(Gx)_i - (x, Gx)].$$

Here the quadratic form

$$(x, Gx) = \Phi$$

represents the value of the differential game, i.e. the mean score of the total population. In this case this will be denoted as the mean fitness of the population.

2.2. Some Examples

To show features of the replicator equation and to clarify the theoretical results in the subsequent sections we will use some suitable examples as often as possible. With these examples we want to demonstrate at the same time the application and results of the implemented program modules.

Example TEST#1: 3-replicator system with symmetric interaction matrix

This example was chosen because of its simplicity, and because a lot of results can be obtained by analytical means.

$$G=\begin{pmatrix} 0 & 1 & 2 \\ 1 & 0 & 4 \\ 2 & 4 & 0 \end{pmatrix}$$

Table 2.2.1 Interaction matrix G for **TEST#1**

We can assign the replicator types 1, 2, and 3 to the vertices, and the interaction coefficients G_{ij} as values to the edges of a graph. The matrix G can then considered as a undirected, edge valued graph as it is depicted in Figure 2.2.1.

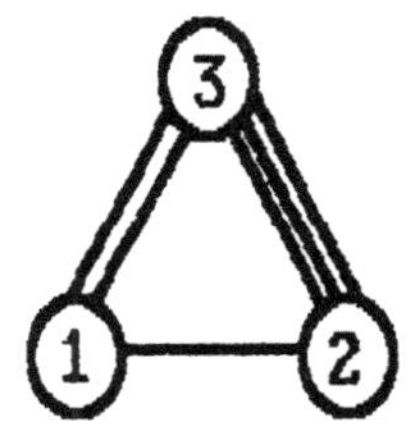

Figure 2.2.1 Graph of example **TEST#1**

Example TEST#31: 10-replicator system with symmetric interaction matrix

Because of the envisioned applications, symmetric matrices are of special importance. The application of a number of program modules will be demonstrated by means of this example.

$$G = \begin{pmatrix} 0 & 1 & 2 & 0 & 0 & 0 & 0 & 0 & 1 & 0 \\ 1 & 0 & 0 & 2 & 1 & 0 & 0 & 0 & 0 & 0 \\ 2 & 0 & 0 & 1 & 0 & 0 & 0 & 1 & 1 & 0 \\ 0 & 2 & 1 & 0 & 0 & 0 & 0 & 3 & 0 & 0 \\ 0 & 1 & 0 & 0 & 0 & 5 & 0 & 0 & 0 & 0 \\ 0 & 0 & 0 & 0 & 5 & 0 & 1 & 0 & 0 & 0 \\ 0 & 0 & 0 & 0 & 0 & 1 & 0 & 1 & 0 & 1 \\ 0 & 0 & 1 & 3 & 0 & 0 & 1 & 0 & 2 & 2 \\ 1 & 0 & 1 & 0 & 0 & 0 & 0 & 2 & 0 & 3 \\ 0 & 0 & 0 & 0 & 0 & 0 & 1 & 2 & 3 & 0 \end{pmatrix}$$

Table 2.2.2 Interaction matrix G for **TEST#31**

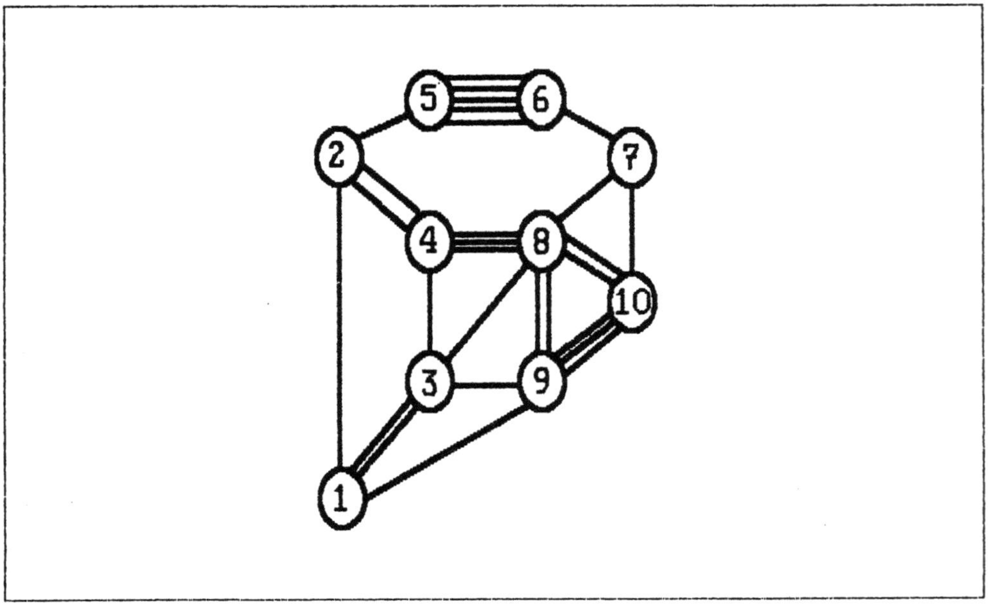

Figure 2.2.2 Graph of example **TEST#31**

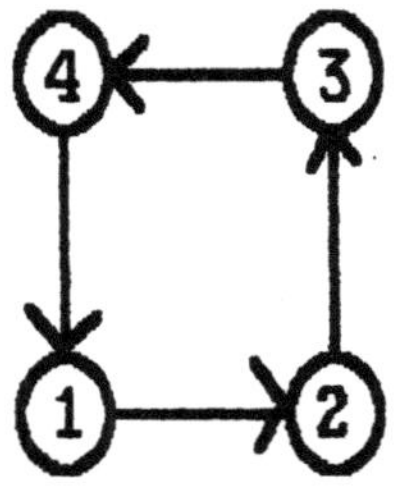

Figure 2.2.3 Graph of example **HYPER1**

Example HYPER1: 4-replicator hypercycle

This example makes clear the influence of feedback to the dynamics of the replicator equation.

$$G = \begin{pmatrix} 0 & 0 & 0 & 1 \\ 1 & 0 & 0 & 0 \\ 0 & 1 & 0 & 0 \\ 0 & 0 & 1 & 0 \end{pmatrix}$$

Table 2.2.3 Interaction matrix G for **HYPER1**

2.3. Replicator Equation and Lotka-Volterra Equation

There exists a strong relation between the Lotka-Volterra equation, the basic equation of theoretical ecology, and the replicator equation, the basic equation of population genetics. This relation was established by Hofbauer (1981).

The Lotka-Volterra equation

$$\dot{y}_i = y_i\left(G_{i0} + \sum_j G_{ij} y_j\right)$$

is defined on the positive orthant

$$R_+^n = \{(y_1, \ldots, y_n) : y_i \geq 0\}.$$

If we add $y_0 = 1$ (as a first integral), and use the transformation

$$x_i = \frac{y_i}{\sum_{j=0}^{n} y_j}$$

then the vector

$$x = (x_0, x_1, \ldots, x_n) \in S_{n+1}$$

is on the simplex S_{n+1}. The inverse transformation is given by

$$y_i = \frac{x_i}{x_0}.$$

Differentiation

$$\dot{x}_i = \frac{\dot{y}_i \sum_j y_j - y_i \sum_j \dot{y}_j}{\left(\sum_j y_j\right)^2}$$

and substitution yields

$$\dot{x}_i = \frac{y_i \sum_j G_{ij} y_j}{\sum_j y_j} - \frac{y_i \sum_k \sum_l y_k G_{kl} y_l}{\left(\sum_j y_j\right)^2}$$

and further

$$\dot{x}_i = \frac{x_i}{x_0} \cdot \frac{\sum_j G_{ij} x_j}{\sum_j x_j} - \frac{\sum_k \sum_l x_k G_{kl} x_l}{\left(\sum_j x_j\right)^2}$$

and

$$\dot{x}_i = \frac{x_i}{x_0} \cdot \left[(Gx)_i - (x, Gx) \right]$$

because of

$$\sum_j x_j = 1 \quad .$$

Up to the factor $1/x_0$ this equation corresponds to the replicator equation

$$\dot{x}_i = x_i \quad \left[(Gx)_i - (x, Gx) \right].$$

But this factor means only a scaling of velocity. Therefore it does not have any influence on the topological structure of the flow on the simplex S_{n+1}.

We will use these topologically equivalent representations in the subsequent sections quite deliberately.

2.4. Fixed Points

For the following analysis stationary points of the replicator equation are of special interest. For these points the condition

$$\dot{x}_i = 0$$

holds. With this condition we get for a fixed point p in int(S_n) the equality

$$(Gp)_i - (p, Gp) = 0.$$

Reformulation yields

$$Gp = (p, Gp)e \text{ with } e = (1, \ldots, 1)' \quad .$$

Inverting G

$$p = (p, Gp) G^{-1} e$$

and summing up all p_i

$$\sum_i p_i = (e, p) = 1 = (p, Gp) \cdot (e, G^{-1}e)$$

and therefore

$$(p, Gp) = \frac{1}{(e, G^{-1}e)}$$

yields the relation for the determination of a fixed point

$$p = \frac{G^{-1}e}{(e, G^{-1}e)}.$$

Taking into account the relation for the adjoint matrix

$$G^{-1} = \frac{adj(G)}{d}$$

we can cancel the determinant d. Thus, the fixed points are only dependent on the adjoint matrix adj(G)

$$p = \frac{adj(G)e}{(e, adj(G)e)}.$$

Fixed points on the faces of the simplex S_n can be computed by deleting the corresponding rows and columns of the interaction matrix G, and applying the same scheme to the reduced fixed point equation.

Of course, by this equation nothing can be said about the stability of the fixed points. It has to be analyzed by means of the corresponding eigenvalues.

2.5. Fixed Points and Saddle Points

Within this section we want to check the existence of a relation between the fixed points of the replicator equation and saddle points of the mean fitness of the total population.

In general a fixed point of the replicator equation is characterized by the relations

$$p_i = 0 \text{ and } (Gp)_i - \Phi \text{ arbitrary}$$

or

$$p_i \neq 0 \text{ and } (Gp)_i - \Phi = 0.$$

We want to show now the identity of fixed points p_i and saddle points x_i of the mean fitness of the total population for real symmetric interaction matrices

$$G_{ij} = G_{ji} \quad , \quad G_{ij} \geq 0$$

with respect to the unit simplex condition. The corresponding Lagrange function has the form

$$L = \Phi - l\left(\sum_i x_i - 1\right) - \sum_{x_i \in \mathrm{bnd}(S_n)} t_i x_i .$$

Then we get

$$\frac{\partial L}{\partial x_i} = 2(Gx)_i - l - t_i = 0$$

and therefore

$$\sum_i x_i \frac{\partial L}{\partial x_i} = 2\Phi - l - \sum_{x_i \in \mathrm{bnd}(S_n)} t_i x_i = 0.$$

Because of

$$\sum_{x_i \in \mathrm{bnd}(S_n)} t_i x_i = 0$$

this yields

$$l = 2\Phi.$$

With the last equation we get the relations for saddle points of the mean fitness by

$$x_i = 0 \text{ and } t_i = 2[(Gx)_i - \Phi] \text{ for } x_i \in \text{bnd}(S_n)$$

or

$$x_i \neq 0 \text{ and } (Gx)_i - \Phi = 0 \text{ for } x_i \notin \text{bnd}(S_n) \quad .$$

But these are exactly the relations for a fixed point of the replicator equation, i.e. saddle points of the mean fitness of the total population are identical with fixed points of the replicator equation with respect to the given conditions. These relations do not hold in general for nonsymmetric interaction matrices.

2.6. Evolutionary Stable Strategies

Within this section we want to emphasize the game theoretic face of the replicator equation. Analyzing fixed points, one may raise quite naturally the question for the existence of a fixed point p in S_n having a mean score which is greater then the score for any other strategy x in $S_n - p$.

For this consideration the following conditions can be formulated:

(i) $(x,Gp)<(p,Gp)$

i.e. if p plays against itself the mean score is greater then the score for a play of x against p. If the score is equal then the relation

(ii) $(x,Gp)=(p,Gp)$ and $(x,Gx)<(p,Gx)$

holds, i.e. if x plays against itself then the score is less than the score for a play of p against x.

Such a strategy p in S_n will be denoted by an evolutionary stable strategy (ESS) (Zeeman (1980)).

Of course, this is a static concept. Now, the following interpretative ideas play a decisive role: An evolutionary development follows the direction of an increase of the mean score, and the mean fitness, respectively, of the total population. At the same time, the development is superposed by random disturbances as e.g. mutations. Because of the afore mentioned conditions and assumptions, during a long term evolutionary process an evolutionary stable strategy will persist.

We want to present here two remarkable theorems (Zeeman (1980)) without proofs:

(i) An ESS is an attractor (in the most simple case a stable fixed point) but not every attractor is an ESS.

(ii) If an ESS p is in the interior of the simplex, i.e. p in int(S_n), then int(S_n) is its basin of attraction, and there exists no other attractor. If x in int(S_n) holds for an attractor then its basin of attraction may be smaller then int(S_n). In this case other attractors may exist in bnd(S_n) (but not in int(S_n)).

For these facts we want to give an example (Zeeman (1980)) with the interaction matrix

$$G = \begin{pmatrix} 0 & 6 & -4 \\ -3 & 0 & 5 \\ -1 & 3 & 0 \end{pmatrix} .$$

The corresponding flow on the simplex S_3 is shown in Figure 2.6.1.

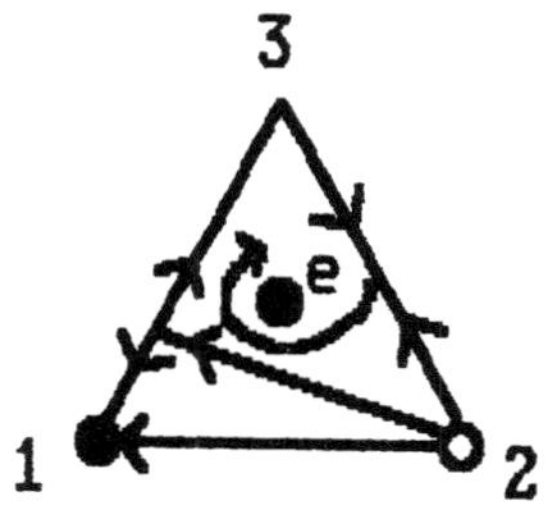

Figure 2.6.1 Flow on the simplex S_3

In this case, the vertex x_1 is an ESS whereas e at the barycentre of the simplex S_3 is not an ESS.

2.7. The Invariance Property

The replicator equation shows a very remarkable invariance property. The application of this property will be very useful for structure formation processes on graphs as we will see later on.

Adding a constant a_j to a column j of the interaction matrix G leaves the replicator equation unchanged (Zeeman (1980)). This can be shown easily.

$$\dot{x}_i = x_i \left[\sum_j (G_{ij} + a_j) x_j - \sum_k \sum_l x_k (G_{kl} + a_l) x_l\right]$$

and therefore

$$\dot{x}_i = x_i \; [(Gx)_i - (x, Gx)] - x_i \left[\sum_j a_j x_j - \sum_k x_k \sum_l a_l x_l\right].$$

Because of

$$\sum_k x_k = 1$$

the second term is zero. This yields the replicator equation

$$\dot{x}_i = x_i \; [(Gx)_i - (x, Gx)] \; .$$

Based on this property, the interaction matrix can be transformed to a matrix with zero diagonal elements. This form of the replicator equation we will denote as the standardized replicator equation.

Here an example: The replicator equation with the interaction matrix

$$G = \begin{pmatrix} 1 & 1 & 2 \\ 1 & 3 & 4 \\ 2 & 4 & 2 \end{pmatrix}$$

can be transformed to the standardized replicator equation with the interaction matrix

$$G=\begin{pmatrix} 0 & -2 & 0 \\ 0 & 0 & 2 \\ 1 & 1 & 0 \end{pmatrix}$$

without changing the flow on the simplex S_3.

2.8. The Fundamental Theorem

For symmetric interaction matrices, the very interesting property can be shown that the mean fitness of the total population cannot decrease in time. This case corresponds to the population genetic model, i.e. to the Fisher-Wright-Haldane model.

This means that

$$\phi \geq 0$$

holds. We want to show this feature (see e.g. Ewens (1979) or Hofbauer & Sigmund (1984)). We have

$$\phi = (x,Gx)^{\cdot} = (\dot{x},Gx)+(x,G\dot{x}) \quad .$$

Because of the symmetry of G

$$\phi = 2\sum_i \dot{x}_i (Gx)_i$$

holds. Replacing dx_i/dt by the right-hand side of the replicator equation yields

$$\phi = 2\sum_i x_i[(Gx)_i-(x,Gx)](Gx)_i$$

and therefore

$$\phi = 2\left[\sum_i x_i (Gx)_i^2-\left(\sum_i x_i (Gx)_i\right)^2\right].$$

Regarding the Jensen inequality

$$\left(\sum_i p_i x_i\right)^a \leq \sum_i p_i x_i^a \text{ for } a>1 \text{ and } (p_1,\ldots,p_n)\in S_n$$

it follows then

$$\dot{\Phi} \geq 0.$$

The equality is valid iff x in S_n is a fixed point. For all other cases, the mean fitness of the gene pool always increases.

The following conclusions are especially interesting:

(i) The increase of the mean fitness of the gene pool is proportional to the variance of the fitness of the replicator types.

(ii) Because of

$$\dot{\Phi} \geq 0$$

the stable fixed points of the replicator equation correspond to the maxima of the quadratic form

$$(x, Gx) \text{ with } x \in S_n \quad ,$$

i.e. the replicator equation can be used to determine the maxima of

$$(x, Gx) \quad .$$

Attractors of the flow of the replicator equation on the simplex S_n are maxima of (x, Gx).

(iii) Because of

$$\Phi > 0 \text{ and } \dot{\Phi} \geq 0 \quad ,$$

the mean fitness of the total population is a global Ljapunov function of the replicator equation for symmetric real nonnegative matrices G.

(iv) The mean fitness f_i of replicator type i is proportional to the gradient of the mean fitness of the gene pool

$$\frac{\partial \Phi}{\partial x_i} \sim (Gx)_i$$

for symmetric matrices G.

The above given property

$$\Phi \geq 0$$

is also valid if some of the x_i are fixed. Without loss of generality we assume that

$$x^1 = (x_1^1, \dots, x_{n_1}^1)' \quad , \quad 1 < n_1 < n$$

are free variables, and the variables

$$x^2 = (x_{n_1+1}^2, \dots, x_n^2)$$

are fixed. The interaction matrix can then be decomposed into the corresponding block matrices

$$G = \begin{pmatrix} G_{11} & G_{12} \\ G_{21} & G_{22} \end{pmatrix} \text{ with } G_{12} = G_{21} \quad .$$

Furthermore we have

$$\sum_j x_j^2 = 1 - m \text{ and } \sum_i x_i^1 = m$$

with $0 < m < 1$. With

$$F_i^1 = (G_{11}x^1)_i + (G_{12}x^2)_i$$

$$F_j^2 = (G_{21}x^1)_j + (G_{22}x^2)_j$$

we get for the mean fitness of the total population the following decomposed equation

$$\Phi = \sum_i x_i^1 F_i^1 + \sum_j x_j^2 F_j^2 = \Phi^1 + \Phi^2 \quad .$$

The replicator equation is then decomposed in the following way

$$\dot{x}_i^1 = x_i^1 (F_i^1 - \Phi)$$

$$0 = x_j^2 (F_j^2 - \Phi)$$

with $i = 1, \dots, n_1$ and $j = n_1 + 1, \dots, n$. From the second equation the mean fitness of the total gene pool

$$\Phi = \frac{\Phi^2}{1-m} \text{ and therefore } \Phi = \frac{\Phi^1}{m} .$$

can be easily computed. Inserting this result into the first part of the replicator equation yields

$$\dot{x}_i^1 = x_i^1 \left(F_i^1 - \frac{\Phi^1}{m} \right) .$$

For

$$\dot{\Phi} = \dot{\Phi}^1 + \dot{\Phi}^2$$

$$= \sum_i \dot{x}_i^1 \left[2(G_{11} x^1)_i + (G_{12} x^2)_i \right] +$$

$$+ \sum_j x_j^2 (G_{21} \dot{x}^1)_j$$

one gets because of $G_{12} = G_{21}$

$$\dot{\Phi} = 2 \sum_i \dot{x}_i^1 \left[(G_{11} x^1)_i + (G_{12} x^2)_i \right]$$

and therefore

$$\dot{\Phi} = 2 \sum_i \dot{x}_i^1 F_i^1 .$$

Inserting the first part of the replicator equation yields

$$\dot{\Phi} = 2 \sum_i x_i^1 F_i^1 (F_i^1 - \Phi)$$

and hence

$$\dot{\Phi} = 2 \left[\sum_i x_i^1 (F_i^1)^2 - \frac{\left(\sum_i x_i^1 F_i^1 \right)^2}{m} \right] .$$

Expanding the right-hand side by m, we have

$$\Phi = 2m\left[\sum_i x_i^1 \frac{(F_i^1)^2}{m} - \left(\sum_i x_i^1 \frac{F_i^1}{m}\right)^2\right] .$$

Applying the Jensen inequality this yields

$$\Phi \geq 0 .$$

More generally, in this case it is possible to show a maximum principle (Sigmund (1985)). Using a suitable definition of a vector product, it is necessary to proof that the trajectories are orthogonal to the constant level set of the mean fitness.

Following Shahshahani (1979), we define for two vectors x and y in the tangent space of S_n at p in int(S_n) the vector product by

$$(x, y)_p = \sum_i \frac{x_i y_i}{p_i} .$$

Let $V(x)$ be a differentiable homogeneous function of degree s and

$$f_i = \frac{\partial V}{\partial x_i}$$

the Euclidean gradients with

$$\frac{\partial f_i}{\partial x_j} = \frac{\partial f_j}{\partial x_i} .$$

Then the trajectories of the replicator equation are orthogonal in the sense of Shahshahani to the constant level set of the potential function $V(x)$ in S_n.

Especially holds that the mean fitness increases with maximal rate in the sense of Shahshahani. The increase is characterized by the equation

$$\Phi = s\dot{V} = s\sum_i \frac{\partial V}{\partial x_i} \cdot \dot{x}_i$$

$$\Phi = s\sum_i \frac{\partial V}{\partial x_i} x_i \left[\frac{\partial V}{\partial x_i} - \sum_j x_j \frac{\partial V}{\partial x_j}\right]$$

$$\Phi = s\left[\sum_i f_i^2 - \left(\sum_i x_i f_i\right)^2\right] \geq 0 .$$

A detailed proof is given by Sigmund (1985).

2.9. The Recurrence Replicator Equation

Up to now, we always assumed that a continuous growth process or reproduction process evolves. Of course, this assumption is not valid for a number of processes, e.g. if the reproduction of a whole population can be seen as a discrete event.

In this case, we get the ratio x'_i of the replicator type i at the total gene pool for the next generation by the recurrence equation

$$x'_i = x_i \frac{(Gx)_i}{(x,Gx)}.$$

This is the recurrence replicator equation with

$$\Phi = (x,Gx)$$

the mean fitness of the total gene pool.

In general, the values of the trajectories of the recurrence replicator equation cannot be considered as sampled values of the trajectories of the continuous replicator equation.

Unfortunately, the invariance property (Section 2.7.) does not hold for the recurrence equation.

Adding a constant a_j to a column j of the interaction matrix G yields

$$x'_i = x_i \frac{\sum_j (G_{ij} + a_j) x_j}{\sum_k \sum_l x_k (G_{kl} + a_l) x_l} .$$

With

$$\sum_k x_k = (e,x) = 1 \text{ and } \sum_j a_j x_j = (a,x)$$

we have the short-hand equation

$$x'_i = x_i \frac{(Gx)_i + (a,x)}{(x,Gx) + (a,x)} \quad .$$

Reformulation yields

$$x'_i = x_i \left[\frac{(Gx)_i}{(x,Gx)} + \left(1 - \frac{x'_i}{x_i}\right) \cdot \frac{(a,x)}{(x,Gx)} \right] \quad .$$

Obviously, the invariance property is not fulfilled in general. But for fixed points

$$x'_i = x_i = p_i \text{ with } p \in \operatorname{int}(S_n)$$

it holds because of

$$1 - \frac{p'_i}{p_i} = 0 \quad .$$

There remains the open question on the influence of adding constants to the columns of the interaction matrix on the qualitative difference of the trajectories of the continuous and the recurrence replicator equation.

Another open question we want to answer in Sections 2.13. and 2.14. concerns possible modifications of the recurrence replicator equation such that the invariance property holds in general.

2.10. Fixed Points

Fixed points of the recurrence replicator equation can be determined in much the same way as in the continuous case. The fixed point condition is given by

$$x'_i = x_i \quad .$$

Therefore we can characterize a fixed point p in $\operatorname{int}(S_n)$ by the equation

$$1 = \frac{(Gp)_i}{(p,Gp)} \text{ or } (Gp)_i - (p,Gp) = 0 \quad .$$

Following the derivation of Section 2.4. we finally get the equation for the computation of a fixed point

$$p = \frac{adj(G)e}{(e, adj(G)e)} \quad .$$

Naturally, the stability and type of the fixed points has to be determined be separate considerations.

As in the case of the continuous replicator equation, the fixed points of the recurrence replicator equation and the saddle points of the mean fitness of the total population are the same for real symmetric interaction matrices.

2.11. The Fundamental Theorem

For the recurrence replicator equation with symmetric interaction matrix it can be also shown that the mean fitness cannot decrease from generation to generation. This means,

$$\Phi' \geq \Phi$$

must hold. We want to show this feature using the elegant proof of Kingman (1961).

$$\Phi' = (x', Gx') = \sum_i \sum_j x'_i G_{ij} x'_j \quad .$$

Substituting the right-hand side of the replicator equation yields

$$\Phi' = \Phi^{-2} \sum_i \sum_j G_{ij} x_i (Gx)_i x_j (Gx)_j$$

$$\Phi' = \Phi^{-2} \sum_i \sum_j \sum_k G_{ij} G_{ik} x_i x_j x_k (Gx)_j .$$

Interchanging i and j

$$\Phi' = \Phi^{-2} \sum_i \sum_j \sum_k G_{ij} G_{ik} x_i x_j x_k (Gx)_k$$

and computing the arithmetic mean of the last two equations gives

$$\Phi' = \frac{\Phi^{-2}}{2} \sum_i \sum_j \sum_k G_{ij} G_{ik} x_i x_j x_k \left[(Gx)_j + (Gx)_k \right].$$

Because of the mean value inequality

$$\frac{(x+y)}{2} \geq (xy)^{\frac{1}{2}}$$

we get the inequality

$$\Phi' \geq \Phi^{-2} \sum_i \sum_j \sum_k G_{ij} G_{ik} x_i x_j x_k \left[(Gx)_j \cdot (Gx)_k \right]^{\frac{1}{2}}$$

or

$$\Phi' \geq \Phi^{-2} \sum_i x_i \left[\sum_j x_j G_{ij} (Gx)_j^{\frac{1}{2}} \right]^2 .$$

Applying the Jensen inequality yields

$$\Phi' \geq \Phi^{-2} \left[\sum_i \sum_j x_j G_{ij} (Gx)_j^{\frac{1}{2}} \right]^2$$

and further

$$\Phi' \geq \Phi^{-2} \left[\sum_j x_j (Gx)_j^{\frac{3}{2}} \right]^2 .$$

Repeated application of the Jensen inequality yields

$$\Phi' \geq \Phi^{-2} \left\{ \left[\sum_j x_j (Gx)_j \right]^{\frac{3}{2}} \right\}^2$$

and therefore

$$\Phi' \geq \Phi^{-2} \left[\sum_j x_j (Gx)_j \right]^3$$

such that finally the relation

$$\Phi' \geq \Phi$$

holds. Here, the equality holds also iff x in S_n is a fixed point.

2.12. The Inverse Replicator Equation

As we already established, the flow of the replicator equation on the simplex S_n runs to the stable fixed points (basins of attraction), and to the maxima of (x,Gx), respectively.

If we assume that the functions $(Gx)_i$ are reciprocal to the growth rates, and the reproductions rates, respectively, i.e.

$$\frac{\dot{x}_i}{x_i} \sim \frac{1}{(Gx)_i} \text{ and } \frac{x'_i}{x_i} \sim \frac{1}{(Gx)_i}$$

then we can define an inverse replicator equation by

$$\dot{x}_i = x_i\left[\frac{1}{(Gx)_i} - \Phi\right] \text{ and } x'_i = \frac{x_i}{(Gx)_i \cdot \Phi}$$

with the normalization term

$$\Phi = \sum_i \frac{x_i}{(Gx)_i}$$

such that the simplex condition $(e,x) = 1$ is valid.

Does there exist any relation between the fixed points of the replicator equation and those of the inverse replicator equation ?

For a fixed point

$$p \in \text{int}(S_n) \text{ with } \dot{p}_i = 0 \text{ or } p'_i = p_i$$

the equation

$$(Gp)_i = \frac{1}{\Phi} \text{ and therefore } Gp = \frac{e}{\Phi}$$

is valid. Summing up over all p_i yields

$$\sum_i p_i = (e,p) = 1 = \frac{(e, G^{-1}e)}{\Phi}$$

and therefore

$$\Phi = (e, G^{-1}e)$$

such that

$$p = \frac{G^{-1}e}{(e, G^{-1}e)} \text{ and } p = \frac{adj(G)e}{(e, adj(G)e)}$$

holds.

Fixed points on the faces of the simplex S_n can be computed by deleting the corresponding rows and columns of the interaction matrix.

With other words, the fixed points of the inverse replicator equation are identical to the fixed points of the replicator equation.

Obviously, the fixed points differ in their stability. It may be assumed that sources, and sinks, respectively, of the flow of the replicator equation on the simplex S_n correspond to sinks, and sources, respectively, of the flow of the inverse replicator equation. Then, the attractors of the inverse replicator equation are minima of the mean fitness of the gene pool (x, Gx).

We want to show that this assumption holds for the population genetic model, i.e. for real, symmetric, nonnegative interaction matrices G, thus obtaining an inverse fundamental theorem.

(i) At first we consider the continuous inverse replicator equation

$$\dot{x}_i = x_i\left[\frac{1}{(Gx)_i} - \Phi\right].$$

We have to show that the mean fitness of the gene pool cannot increase, i.e.

$$\dot{\Phi} \leq 0$$

must hold. Because of the symmetry of G we have

$$\dot{\Phi} = 2\sum_i \dot{x}_i (Gx)_i$$

Inserting the right-hand side of the inverse replicator equation yields

$$\dot{\Phi} = 2\sum_i x_i (Gx)_i\left[\frac{1}{(Gx)_i} - \Phi\right]$$

and therefore

$$\dot{\Phi} = 2\left[\sum_i x_i - \hat{\Phi}\sum_i x_i (Gx)_i\right] \text{ or } \dot{\Phi} = 2[1 - \hat{\Phi}\Phi] \quad .$$

Because of the relation between the arithmetic and the geometric mean for

$$(Gx)_i \geq 0 \text{ and } (e,x) = 1$$

we have

$$\sum_i x_i (Gx)_i \geq \frac{1}{\sum_i \frac{x_i}{(Gx)_i}} \text{ or } \hat{\Phi} \cdot \Phi \geq 1$$

and finally the inequality

$$\dot{\Phi} \leq 0 \quad .$$

(ii) For the recurrence inverse replicator equation

$$x'_i = \frac{x_i}{(Gx)_i \hat{\Phi}}$$

the change of the mean fitness of the gene pool is given by

$$\Phi' = (x', Gx') \text{ or } \Phi' = \hat{\Phi}^{-2} \sum_i \frac{x_i}{(Gx)_i} \sum_j G_{ij} \frac{x_j}{(Gx)_j} \quad .$$

Recasting yields

$$\Phi' = \hat{\Phi}^{-2} \sum_i x_i \sum_j G_{ij} x_j \left\{ \left[\frac{1}{(Gx)_i \cdot (Gx)_j} \right]^{\frac{1}{2}} \right\}^2 \quad .$$

Because of the relation between the arithmetic and the geometric mean for nonnegative $(Gx)_i$ we have

$$\sum_i x_i \sum_j G_{ij} x_j \left\{ \left[\frac{1}{(Gx)_i \cdot (Gx)_j} \right]^{\frac{1}{2}} \right\}^2 \leq \frac{1}{4} \sum_i x_i \sum_j G_{ij} x_j \left[\frac{1}{(Gx)_i} + \frac{1}{(Gx)_j} \right]^2 \quad .$$

Expanding the quadratic term on the right-hand side and taking into account the symmetry of G yields

$$\frac{1}{4}\sum_i x_i \sum_j G_{ij} x_j \left[\frac{1}{(Gx)_i} + \frac{1}{(Gx)_j}\right]^2 =$$

$$= \frac{1}{4}\sum_i x_i \sum_j G_{ij} x_j \left[\frac{1}{(Gx)_i^2} + \frac{2}{(Gx)_i \cdot (Gx)_j} + \frac{1}{(Gx)_j^2}\right]$$

$$= \frac{1}{4}\left[2\sum_i \frac{x_i}{(Gx)_i^2}\sum_j G_{ij} x_j + 2\sum_i \frac{x_i}{(Gx)_i}\sum_j G_{ij}\frac{x_j}{(Gx)_j}\right] .$$

Hence, we have

$$\sum_i \frac{x_i}{(Gx)_i}\sum_j G_{ij}\frac{x_j}{(Gx)_j} \le \bar{\Phi} .$$

Inserting this relation into the equation for the mean fitness of the next generation yields

$$\Phi' \le \frac{1}{\bar{\Phi}} .$$

Because of the inequality

$$\bar{\Phi} \cdot \Phi \ge 1$$

we finally have

$$\Phi' \le \Phi .$$

But this means, stable fixed points of the continuous and the recurrence inverse replicator equation correspond to the minima of the mean fitness of the gene pool.

Hence, we have the possibility to determine sources of the flow of the replicator equation on the simplex S_n, and therefore to determine minima of the function (x, Gx).

2.13. The Generalized Replicator Equation

Both, the continuous and the recurrence replicator equation can be generalized. For this reason we consider a strictly monotone increasing function g with

$$g(f_i)>g(f_j) \text{ for } f_i>f_j \quad .$$

Let g^{-1} be the unique inverse function. Functions of such type play an important role in connection with generalized superposition principles (Kolmogoroff (1930), Gössel (1972), Peschel & Mende (1983)).

Mapping the fitness functions f_i of replicator types i by means of the function g yields the generalized replicator equation

$$\dot{x}_i=x_i\left[g(f_i)-\sum_i x_i g(f_i)\right] \text{ and } x'_i=x_i\frac{g(f_i)}{\sum_i x_i g(f_i)}.$$

Of course, a number of questions arise concerning the features of this generalized equation.

Since we are especially interested in fixed points, we want to analyze them. A fixed point of the generalized replicator equation is characterized by

$$p_i=0 \text{ and } g(f_i)-\sum_i p_i g(f_i) \text{ arbitrary}$$

or

$$p_i\neq 0 \text{ and } g(f_i)-\sum_i p_i g(f_i)=0 \quad .$$

Considering fixed points p in $\text{int}(S_n)$ we have

$$g(f_i)=\sum_i p_i g(f_i)$$

and therefore

$$f_i=g^{-1}\left(\sum_i p_i g(f_i)\right).$$

The right-hand side of this equation is a generalization of the usual linear superposition which was introduced by Kolmogoroff in connection with mean value operators. It is especially interesting that obviously

$$\Phi=\sum_i p_i f_i = g^{-1}\left(\sum_i p_i g(f_i)\right)$$

holds for a fixed point p in S_n.

Hence, the mean fitness of the total population has the same values at fixed points of the replicator equation and of the generalized replicator equation.

For any state x in S_n of the generalized replicator equation, the Jensen inequality

$$\sum_i x_i g(f_i) \geq g\left(\sum_i x_i f_i\right)$$

holds because of the strict monotony of g which implies its convexity.

Let us now consider linear fitness functions $f_i = (Gx)_i$. For a fixed point p in int(S_n) we have the equation

$$Gp = g^{-1}\left(\sum_i p_i g((Gp)_i)\right)e$$

and therefore

$$p = g^{-1}\left(\sum_i p_i g((Gp)_i)\right)G^{-1}e.$$

Summing up all p_i yields

$$\sum_i p_i = (e,p) = 1 = g^{-1}\left(\sum_i g((Gp)_i)\right)(e, G^{-1}e) \;.$$

Inserting this relation into the previous equation we have finally

$$p = \frac{G^{-1}e}{(e, G^{-1}e)} \;.$$

Hence, the position of fixed points of the replicator equation is invariant with respect to the application of a strictly monotone increasing mapping of the fitness functions f_i of the replicator types i.

Now we want to consider the invariance of the generalized replicator equation with respect to changes of the interaction coefficients between the replicator types (see Sections 2.7. and 2.9.).

Once again we use linear fitness functions f_i, and analyze the influence of adding a constant a_j to a column j of the interaction matrix.

For the continuous generalized replicator equation we have

$$\dot{x}_i = x_i\left[g(f_i + (a,x)) - \sum_l x_l g(f_l + (a,x))\right] \text{ with } (a,x) = \sum_j a_j x_j.$$

If the function g has the property

$$g(x+y) = g(x) + g(y)$$

then the continuous generalized replicator equation is invariant with respect to the mentioned influence on the interaction matrix.

For the recurrence generalized replicator equation

$$x'_i = x_i \frac{g(f_i + (a,x))}{\sum_l x_l g(f_l + (a,x))}$$

similar results will be obtained. If

$$g(x+y) = g(x) \cdot g(y)$$

holds then the recurrence generalized replicator equation is invariant with respect to adding a constant to a column of the interaction matrix.

This holds e.g. for the function

$$g(x) = e^x \quad .$$

For this function the recurrence generalized replicator equation is given by

$$x'_i = x_i \frac{e^{f_i}}{\sum_l x_l e^{f_l}}$$

with $f_i = (Gx)_i$.

This replicator equation has the following features in common with the continuous replicator equation·

(i) Simplex invariance

(ii) Invariance with respect to adding a constant to a column of the interaction matrix

(iii) The position of fixed points of both equations is identical.

A detailed analysis follows in the next sections.

2.14. The Natural Recurrence Replicator Equation

Within this section we introduce the natural recurrence replicator equation (Voigt (1988)). This equation is characterized by the same important invariance features as the continuous replicator equation.

To get a self-contained treatment we give a short recall of these features.

Feature 1

With respect to the Lotka-Volterra equation

$$\dot{y}_i = y_i\left(G_{in} + \sum_j G_{ij} y_j\right) \quad , \quad i = 1,\dots,n-1$$

the positive orthant

$$R_+^{n-1} = \{(y_1,\dots,y_{n-1}) : y_i \geq 0\}$$

is invariant.

Feature 2

With respect to the replicator equation

$$\dot{x}_i = x_i\left(\sum_j G_{ij} x_j - \Phi\right) \quad , \quad i = 1,\dots,n$$

the unit simplex

$$S_n=\left\{(x_1,\dots,x_n):x_i\geq 0,\sum_i x_i=1\right\}$$

is invariant.

Feature 3

The Lotka-Volterra equation is equivalent to the replicator equation, and vice versa, by the change of coordinates

$$y_i=\frac{x_i}{x_n} \text{ or } x_i=\frac{y_i}{1+\sum_j y_j}.$$

Feature 4

Adding any constant a_j to a column j of the interaction matrix G

$$G_{ij}:=G_{ij}+a_j$$

leaves the replicator equation invariant.

Feature 4 plays an important role in dynamic game theory (Zeeman (1980)).

Unfortunately, all features do not hold in general for the recurrence Lotka-Volterra equation

$$y'_i=y_i\left(G_{in}+\sum_j G_{ij}y_j\right)$$

and the recurrence replicator equation

$$x'_i=x_i\frac{\sum_j G_{ij}x_j}{\Phi}.$$

Especially

R_+^{n-1}-invariance and S_n-invariance

are violated.

To ensure these features we use a nonlinear mapping g of the fitness functions f_i mentioned in the previous section with the following properties

(i) g is strictly monotone increasing.

(ii) For any two elements x, y the relation

$$g(x+y)=g(x)\cdot g(y)$$

holds.

Property (i) ensures the character of a fitness function, and property (ii) supports feature 4. Because of the strict monotony there exists a unique inverse thus ensuring fixed point relations.

Obviously the mapping

$$g(x)=a^x \quad , \quad a>1$$

or

$$g(x)=e^{D\cdot x} \quad , \quad D>0$$

is of such type. For this mapping we have the natural recurrence replicator equation

$$x'_i=x_i\frac{a^{\sum_j c_{ij}x_j}}{\Phi} \quad , \quad a>1 \text{ with } \Phi=\sum_i x_i a^{\sum_j c_{ij}x_j}$$

and the natural recurrence Lotka-Volterra equation

$$y'_i=y_i a^{\left(c_{in}+\sum_j c_{ij}y_j\right)} \quad , \quad a>1 \ .$$

We want to show that features 1 to 4 hold for these equations.

Feature 1

Assume

$$y_i\in R_+^{n-1} \ .$$

Since

$$a^{\left(G_{in}+\sum_j G_{ij}y_j\right)}>0 \text{ for } a>1 \text{ and any } G_{ij}\ ,\ G_{in}\ ,$$

$$y'_i \in R_+^{n-1}$$

holds.

Feature 2

Assume

$$x_i \in S_n\ .$$

Since

$$a^{\sum_j G_{ij}x_j}>0 \text{ for } a>1 \text{ and any } G_{ij}\ ,$$

$$\Phi>0$$

and therefore

$$x'_i \in S_n$$

holds.

Feature 3

For the natural recurrence Lotka-Volterra equation let us assume the change of coordinates

$$x_i=\frac{y_i}{\sum_j y_j}\ ,\quad i,j=1,\ldots,n$$

with $y_n = 1$. Then we have finally

$$x'_i=x_i\frac{a^{\frac{1}{x_n}\sum_j G_{ij}}}{\sum_j x_j a^{\frac{1}{x_n}\sum_l G_{jl}x_l}}\ .$$

With

$$a:=a^{1/x_n}$$

this yields the natural recurrence replicator equation.

Feature 4

Adding a constant a_j to a column of the interaction matrix of the natural recurrence replicator equation, we have

$$x'_i = x_i \frac{a^{\sum_j (c_{ij}+a_j)x_j}}{\sum_l x_l a^{\sum_i (c_{li}+a_i)x_i}} .$$

This yields

$$x'_i = x_i \frac{a^{\sum_j c_{ij}x_j} a^{\sum_j a_j x_j}}{a^{\sum_j a_j x_j} \sum_l x_l a^{\sum_i c_{li}x_i}}$$

and therefore

$$x'_i = x_i \frac{a^{\sum_j c_{ij}x_j}}{\sum_l x_l a^{\sum_i c_{li}x_i}} .$$

Hence, the natural recurrence replicator equation is invariant with respect to such transformations just like the continuous replicator equation.

The crucial point of this derivation is the rescaling

$$a := a^{\frac{1}{x_n}}$$

in feature 3 as an analogue to time transformations in the time continuous equation. Though time scaling with differential equations is usually a noncritical transformation, the above given rescaling may lead to dynamical structures such as limit cycles or even chaotic behaviour.

What is possible to say is that the positions of fixed points of the continuous replicator equation and the natural recurrence replicator equation are identical (at most in the interior of the unit simplex int(S_n)):

A fixed point z in int(S_n) of the continuous replicator equation is given by

$$z=\frac{adj(G)e}{(e,adj(G)e)}.$$

With

$$z_i=x'_i=x_i$$

we have for the natural recurrence replicator equation

$$\sum_j z_j a^{\sum_i G_{ji}x_i}=a^{\sum_j G_{ij}x_j}$$

and therefore

$$\sum_j G_{ij}z_j=\log_a\left(\sum_j a^{\sum_i G_{ji}x_i}\right)=r$$

or in comprised form

$$Gz=re.$$

This yields

$$z=rG^{-1}e$$

or because of the simplex condition

$$\sum_i z_i=(e,z)=1=r\left(e,G^{-1}e\right) \text{ or } r=\frac{1}{\left(e,G^{-1}e\right)}.$$

Therefore the fixed point relation

$$z=\frac{G^{-1}e}{\left(e,G^{-1}e\right)}=\frac{adj(G)e}{(e,adj(G)e)}$$

holds for the natural recurrence replicator equation. This equation is the same as in the time continuous case.

Finally we want to clarify the relation of the natural recurrence Lotka-Volterra equation to the direct iteration method (Peschel, Mende & Breitenecker (1987)).

Let us consider the natural recurrence Lotka-Volterra equation

$$y'_i = y_i a^{\sum_j G_{ij} y_j} .$$

This may be written as

$$y'_i = y_i \frac{\prod_{j, G_{ij}>0} a^{G_{ij} y_j}}{\prod_{j, G_{ij}<0} a^{-G_{ij} y_j}} .$$

Linear approximation of the single terms

$$a^{G_{ij} y_j} = 1 + \ln(a) G_{ij} y_j$$

and $\ln(a) = D$ yields the direct iteration representation of the Lotka-Volterra equation

$$y'_i = y_i \frac{\prod_{j, G_{ij}>0} (1 + D G_{ij} y_j)}{\prod_{j, G_{ij}<0} (1 - D G_{ij} y_j)} .$$

3. The Controlled Replicator Equation

With the classical example of fishing in the Adriatic Sea, mentioned in the introduction, we roughly considered the influence of fishing intensity as a control variable. In the following we will analyze a goal oriented influencing of elements of the interaction matrix as a control problem. We are interested in the structures of the total population, and the underlying graphs, respectively, which are possible to generate.

3.1. Nonhomogeneous Replication Rate Control

Based on the invariance property (see Section 2.7.) of the replicator equation the interaction matrix can be always mapped to the standardized form

$$G = \begin{pmatrix} 0 & * & * & * & * & * & * \\ * & 0 & * & * & * & * & * \\ * & * & . & * & * & * & * \\ * & * & * & . & * & * & * \\ * & * & * & * & . & * & * \\ * & * & * & * & * & 0 & * \\ * & * & * & * & * & * & 0 \end{pmatrix}.$$

Here, the diagonal elements are identical to zero, $G_{ii} = 0$, and the off-diagonal elements G_{ij} have arbitrary values.

The diagonal elements represent the self-reproduction rates or self-replication rates. It is quite natural to introduce now control variables u_i to influence the self-replication rates (Voigt (1982,1985)). This yields the interaction matrix

$$\bar{G} = \begin{pmatrix} u_1 & * & * & * & * \\ * & u_2 & * & * & * \\ * & * & . & * & * \\ * & * & * & . & * \\ * & * & * & * & u_n \end{pmatrix}$$

or in short-hand notation

$$\bar{G} = G + diag(u_i)_1^n = G + D_u.$$

We have then the controlled replicator equation

$$\dot{x}_i = x_i[(Gx)_i - (x, Gx)] + x_i\left[u_i x_i - \sum_j u_j x_j^2\right].$$

By means of the variables u_i, the self-replication rates can be controlled completely independent. Thus the controls are acting completely selective on the different replicator types. Using not completely selective controls, this feature has to be taken into account.

The first problem we want to consider is the following: Is it possible to reach a stable final state x in int(S_n) starting at an arbitrary initial state z in int(S_n) with a fixed control vector

$$u = (u_1, \ldots, u_n) \quad ?$$

How to determine the corresponding values of the control vector ?

We want to show that the construction of u with the desired effect is always possible. For the proof we use the conditions for evolutionary stable strategies, and the feature that an ESS x in int(S_n) has S_n as the basin of attraction.

First we generate an arbitrary fixed point x in int(S_n) with $dx_i/dt = 0$. This reason for we write the controlled replicator equation in vector notation

$$\dot{x} = D_x[(G + D_u)x - (x, (G + D_u)x)e]$$

with

$$D_x = diag(x_i)_1^n \text{ and } D_u = diag(u_i)_1^n \quad .$$

For $dx/dt = 0$ we have

$$(G + D_u)x = (x, (G + D_u)x)e \quad .$$

Recasting yields

$$D_x x-(x,D_x x)e=(x,Gx)e-Gx$$

and furthermore

$$(E-X)D_x u=(x,Gx)e-Gx$$

with

$$X=\begin{pmatrix} x_1 & x_2 & \dots & x_n \\ x_1 & x_2 & \dots & x_n \\ . & . & \dots & . \\ x_1 & x_2 & \dots & x_n \end{pmatrix}$$

and E the unit matrix. With

$$rank(E-X)=n-1$$

because of $(e,x) = 1$, this yields

(i) $x \in \text{int}(S_n)$

$$rank D_x = n$$

and therefore

$$rank(E-X)D_x = n-1$$

(ii) $x \in \text{bnd}(S_n)$

$$rank D_x = k < n$$

and therefore because of the Sylvester inequality

$$k-1 < rank(E-X)D_x < k \quad .$$

For x in int(S_n) we get the control u as the solution of a system of linear equations with one degree of freedom. Applying to the system of equations

$$\begin{pmatrix} 1-x_1 & -x_2 & \dots & -x_n \\ -x_1 & 1-x_2 & \dots & -x_n \\ \dots & \dots & \dots & \dots \\ -x_1 & -x_2 & \dots & 1-x_n \end{pmatrix} D_x u = \begin{pmatrix} (x,Gx)-(Gx)_1 \\ (x,Gx)-(Gx)_2 \\ \dots \\ (x,Gx)-(Gx)_n \end{pmatrix}$$

a Gauss elimination yields

$$\begin{pmatrix} 1 & 0 & 0 & \dots & -1 \\ -1 & 1 & 0 & \dots & 0 \\ 0 & -1 & 1 & \dots & 0 \\ . & . & . & \dots & . \\ 0 & 0 & 0 & \dots & 0 \end{pmatrix} D_x u = \begin{pmatrix} (Gx)_n - (Gx)_1 \\ (Gx)_n - (Gx)_2 \\ (Gx)_n - (Gx)_3 \\ \dots \\ 0 \end{pmatrix} .$$

u_n can be considered as a free control variable. For given u_i we have the single equations

$$u_i = \frac{1}{x_i}[x_n u_n + (Gx)_n - (Gx)_i] \text{ for } i = 1,\dots,n-1 \quad .$$

Up to now, nothing is known about the stability and the basin of attraction of the fixed point x in int(S_n), generated by the controls u_i. For this reason we use a theorem from Section 2.6.

We show now that for a suitable free control variable u_n the fixed point x in int(S_n) follows the conditions of an ESS. We do this by two steps.

(i) $(z,\bar{G}x) < (x,\bar{G}x)$

We write this condition as

$$f(z) = (x,\bar{G}x) - (z,\bar{G}x) > 0$$

or in expanded form

$$f(z) = (x,(G + D_u)x) - (z,(G + D_u)x) > 0 \quad .$$

By transformation we get the quadratic form

$$f(z) = (x - z,(G + D_u)x) = \sum_i (x_i - z_i)((G + D_u)x)_i \quad .$$

Because x in int(S_n) is a fixed point, the elements of the column vector

$$(G + D_u)x$$

are independent of i, i.e.

$$((G + D_u)x)_i = ((G + D_u)x)_. \quad .$$

Because

$$(e,x) = (e,z) = 1$$

this yields

$$f(z)=0 \text{ for } z,x\in \operatorname{int}(S_n) \quad ,$$

and therefore (i) does not hold.

(ii) $(z,\overline{G}z)<(x,\overline{G}z)$

This condition can be written as

$$g(z)=(x,\overline{G}z)-(z,\overline{G}z)>0$$

or in expanded form

$$g(z)=(x,(G+D_u)z)-(z,(G+D_u)z)>0.$$

Thus we have the quadratic form

$$g(z)=(x-z,(G+D_u)z).$$

Inserting the solutions

$$u_i=\frac{1}{x_i}[x_n u_n+(Gx)_n-(Gx)_i]$$

yields

$$g(z)=-x_n u_n(x-z,D_x^{-1}(x-z))-(x-z,D_x^{-1}(D_xG+(Gx)_nE-D_{(Gx)})(x-z)).$$

With the substitutions

$$v=x-z$$

$$l=-x_n u_n$$

$$A=D_x^{-1}$$

$$B=D_x^{-1}((D_xG+(Gx)_nE-D_{(Gx)})$$

$$D_{(Gx)}=diag((Gx)_i)_1^n$$

we get finally the matrix bunch

$$g(v) = (v,(lA - B)v).$$

u_n, and therefore l can be chosen always such that

$$g(v) > 0$$

is a positive definite quadratic form.

Then x in int(S_n) is an ESS with int(S_n) as the basin of attraction. Therefore for any initial state z in int(S_n) the final stable state will always be x in int(S_n).

Of course, nothing is known about the path of the state transition. If certain functionals has to be taken into account for the state transition, then the final state x can be reached via a sequence of control values $\{u^k\}$ and therefore via a sequence of intermediate states $\{x^k\}$. In this case, the computation of the control values can be done as given above.

Now the question has to be raised about further features of fixed points p in int(S_n) generated by the control vector u. What kind of dependency of the eigenvalues on the free control variable u_n does exist?

Here we can state the following theorem:
The eigenvalues of a fixed point p in int(S_n) generated by the control u with the free control variable u_n are in the real parts linearly dependent on u_n.

We want to show this feature and write the replicator equation in the form

$$\dot{x}_i = f_i(x) = x_i[(Gx)_i - (x,Gx)] \quad .$$

For a fixed point p in int(S_n) we have the elements of the Jacobian by

$$\frac{\partial f_i}{\partial x_k}\Big|_p = p_i\Big[G_{ik} - G_{in} - \sum_j (G_{jk} - G_{jn})p_j\Big] \quad , \quad i = 1,\dots,n \quad .$$

Inserting as diagonal elements of the matrix G the control variables

$$G_{ii} = u_i = \frac{1}{p_i}[p_n u + (Gp)_n - (Gp)_i]$$

with $u = u_n$ then the diagonal elements of the Jacobian are only depending on u

$$\frac{\partial f_i}{\partial x_k}\Big|_p = \begin{cases} a_{ik} & \text{for } i \neq k \\ p_n u + a_{ii} & \text{for } i = k \end{cases}$$

or in matrix notation

$$\left(\frac{\partial f_i}{\partial x_k}\right)_p = p_n u E + A \quad .$$

The corresponding eigenvalue equation is then given by

$$det(lE - p_n uE - A) = det((l - p_n u)E - A) = 0$$

or with

$$\mu = l - p_n u$$

by

$$det(\mu E - A) = 0 \quad .$$

This yields for the eigenvalues

$$l_i = (Re(\mu_i) + p_n u) + Im(\mu_i) \quad .$$

Because $0 < p_n < 1$ and u real only the real parts of l_i will be influenced by the control $u = u_n$.

Generally, and especially because of the subsequent details we consider symmetric matrices G.

In this case we have the following assertion:

For real symmetric matrices G the eigenvalues of a fixed point p in int(S_n) are real.

To show this we consider for the replicator equation

$$\dot{x}_i = f_i(x) = x_i[(Gx)_i - (x, Gx)]$$

the Jacobian in the following form

$$\frac{\partial f_i}{\partial x_k} = \begin{cases} x_i[G_{ik} - 2(Gx)_k] & \text{for } i \neq k \\ (Gx)_i - (x, Gx) + x_i[G_{ik} - 2(Gx)_k] & \text{for } i = k \end{cases} .$$

For a fixed point p in int(S_n) we have

$$\frac{\partial f_i}{\partial x_k}\Big|_p = p_i[G_{ik} - 2(Gp)_k] \quad .$$

Because of the fixed point relation

$$(Gp)_i = (p, Gp) \text{ for } i = 1, \dots, n$$

this yields

$$\frac{\partial f_i}{\partial x_k}\Big|_p = p_i[G_{ik} - 2(p, Gp)] \quad .$$

The matrix

$$[\overline{G}_{ik}] = [G_{ik} - 2(p, Gp)]$$

is also real and symmetric. Hence the Jacobian has the following form

$$\left(\frac{\partial f_i}{\partial x_k}\right)_p = J = D_p\overline{G} \text{ with } D_p = diag\{p_i\}_1^n \quad .$$

By means of the similarity transformation

$$D_{\sqrt{p}} = diag\{\sqrt{p_i}\}_1^n$$

J can be transformed to the real symmetric matrix

$$\mathcal{J} = D_{\frac{1}{\sqrt{p}}} J D_{\sqrt{p}} = D_{\sqrt{p}}\overline{G}D_{\sqrt{p}} .$$

Hence the eigenvalues of both matrices are real.

3.2. Homogeneous Replication Rate Control

Now we want to analyze replication rate controls

$$u = u_i \quad , \quad i = 1, \dots, n$$

in detail.

In this case the interaction matrix has the form

$$\bar{G} = G + uE.$$

Generally, for a fixed point $p(u)$ in int(S_n) the relations

$$p(u) = \frac{\bar{G}^{-1}(u)e}{(e, \bar{G}^{-1}(u)e)}$$

or

$$p(u) = \frac{adj(\bar{G}(u))e}{(e, adj(\bar{G}(u))e)}$$

hold. The adjoint matrix can be written as a matrix polynomial

$$adj(\bar{G}(u)) = \sum_{i=0}^{n-1} B_i u^{n-1-i}$$

with constant matrices B_i. The matrices B_i can be computed e.g. by the following scheme (Faddejeva & Faddejev (1964))

$$\begin{aligned}
B_0 &= E, d_1 = tr(G)\\
B_1 &= -B_0G + d_1E, d_2 = \frac{1}{2}tr(B_1G)\\
B_2 &= -B_1G + d_2E, d_3 = \frac{1}{3}tr(B_1G)\\
&\dots\\
B_{n-1} &= -B_{n-2}G + d_nE, d_n = \frac{1}{n}tr(B_{n-1}G)\\
0 &= -B_{n-1}G + d_nE
\end{aligned}$$

The last equation is not necessary but can be used for the estimation of rounding errors.

The components of a fixed point $p_i(u)$ are then be given by

$$p_i(u) = \frac{\sum_j p_{ij} u^{n-j}}{\sum_i p_i(u)} \quad .$$

For the example **TEST#1** the coefficients p_{ij} and the real zeros of the polynomials

$$\sum_j p_{ij} u^{n-j}$$

are shown in Table 3.2.1.

COEFFICIENTS OF COMPONENT POLYNOMIALS

1.000000E+00	-3.000000E+00	-4.000000E+00
1.000000E+00	-5.000000E+00	6.000000E+00
1.000[illegible]E+00	-6.000000E+00	5.000000E+00

REAL ZEROS OF COMPONENT POLYNOMIALS

-1.000000E+00	4.000000E+00
2.000000E+00	3.000000E+00
1.000000E+00	5.000000E+00

Table 3.2.1 Coefficients and real zeros of the fixed point polynomials

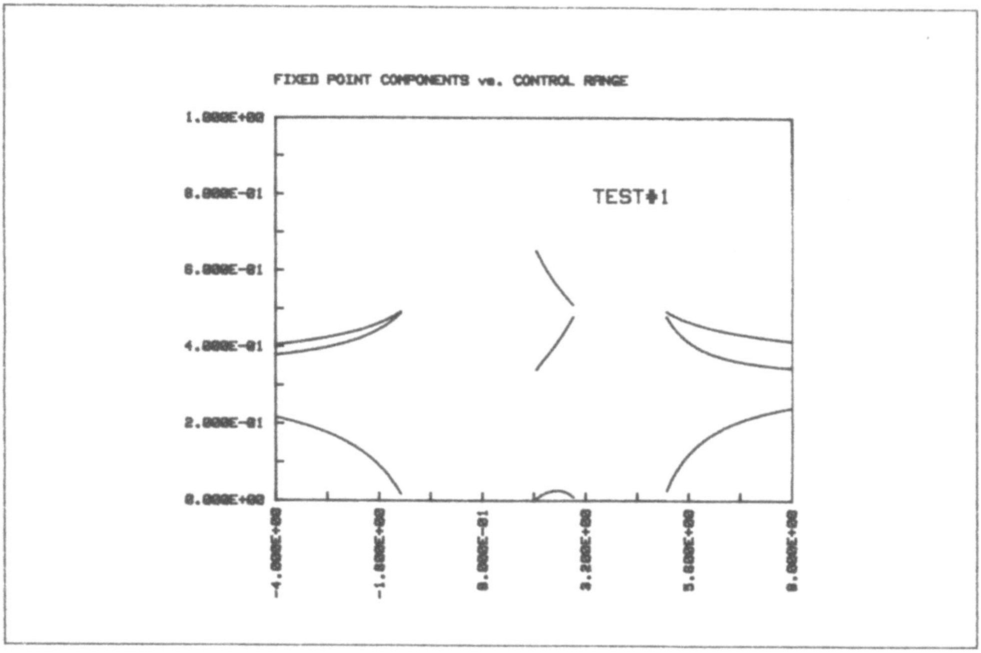

Figure 3.2.1 Graph of the fixed point components $p_i(u)$ for **TEST#1**

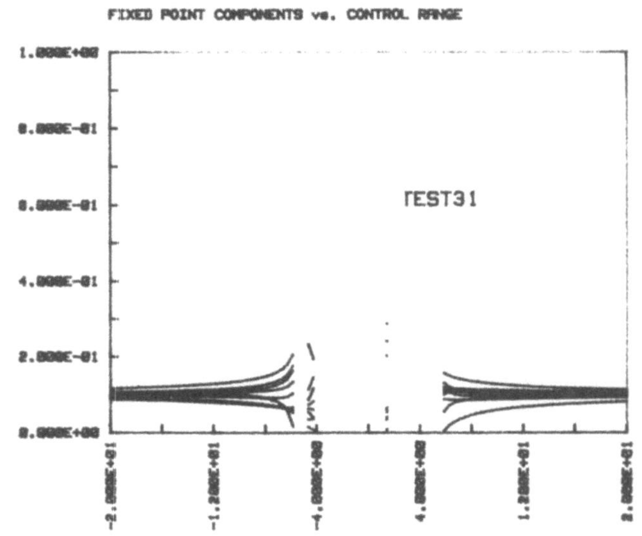

Figure 3.2.2 Graph of the fixed point components $p_i(u)$ **for TEST#31** ($-20 < u < 20$)

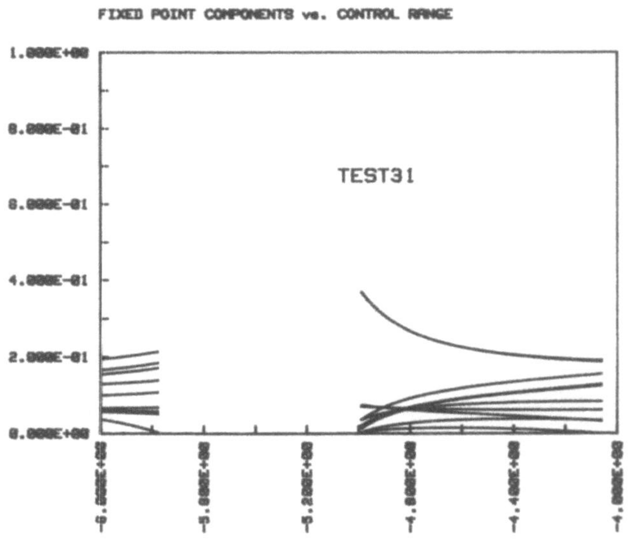

Figure 3.2.3 Graph of the fixed point components $p_i(u)$ **for TEST#31** ($-6 < u < -4$)

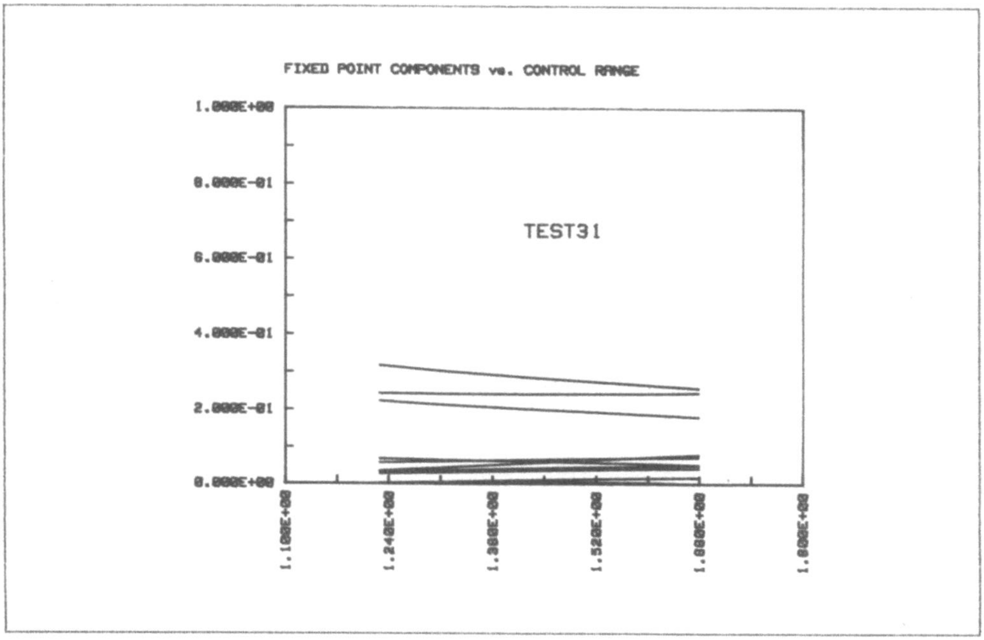

Figure 3.2.4 Graph of the fixed point components $p_i(u)$ for **TEST#31** $(1.1 < u < 1.8)$

```
JACOBIAN

-1.297297E+00 -2.162162E-01
-1.135135E+00 -3.027027E+00

FIXED POINT FOR DIAGONAL CONTROL VALUE = -4.000000E+00

 2.162162E-01   3.783784E-01   4.054054E-01

EIGENVALUES OF THE REPLICATOR EQUATION FOR ABOVE FIXED POINT
      REAL          IMAGINARY       INDICATOR

-3.158870E+00   0.000000E+00   1.000000E+00
-1.165455E+00   0.000000E+00   1.000000E+00
```

```
JACOBIAN

-1.500000E-11 -2.916667E-12
-1.500000E+00 -2.500000E+00

FIXED POINT FOR DIAGONAL CONTROL VALUE = -1.000000E+00

 2.916667E-12  5.000000E-01  5.000000E-01

EIGENVALUES OF THE REPLICATOR EQUATION FOR ABOVE FIXED POINT
     REAL         IMAGINARY      INDICATOR

-2.500000E+00  0.000000E+00  1.000000E+00
-1.250000E-11  0.000000E+00  1.000000E+00
```

```
JACOBIAN

 2.837838E-01 -5.675676E-01
-8.108108E-02 -4.054054E-02

FIXED POINT FOR DIAGONAL CONTROL VALUE = 2.500000E+00

 5.675676E-01  2.702703E-02  4.054054E-01

EIGENVALUES OF THE REPLICATOR EQUATION FOR ABOVE FIXED POINT
     REAL         IMAGINARY      INDICATOR

-1.472939E-01  0.000000E+00  1.000000E+00
 3.905371E-01  0.000000E+00  1.000000E+00
```

```
JACOBIAN

 5.000000E-01 -5.000000E-01
-7.500000E-12  7.500000E-12

FIXED POINT FOR DIAGONAL CONTROL VALUE = 3.000000E+00

 5.000000E-01  2.500000E-12  5.000000E-01

EIGENVALUES OF THE REPLICATOR EQUATION FOR ABOVE FIXED POINT
     REAL         IMAGINARY      INDICATOR

 0.000000E+00  0.000000E+00  1.000000E+00
 5.000000E-01  0.000000E+00  1.000000E+00
```

```
JACOBIAN

 1.806452E+00 -4.516129E-01
-1.161290E+00  7.741935E-01

FIXED POINT FOR DIAGONAL CONTROL VALUE = 6.000000E+00

 4.516129E-01  3.870968E-01  1.612903E-01

EIGENVALUES OF THE REPLICATOR EQUATION FOR ABOVE FIXED POINT
     REAL         IMAGINARY      INDICATOR

4.010291E-01  0.000000E+00  1.000000E+00
2.179616E+00  0.000000E+00  1.000000E+00
```

Table 3.2.2 Jacobian, fixed points and eigenvalues for different values of u for example **TEST#1**

The graph of the fixed point components $p_i(u)$ is shown in Figure 3.2.1. For the example **TEST#31** figs. 3.2.2. to 3.2.4. show the graphs of the fixed point components for different ranges of the control u. The figures show quite clear the intervals where no fixed point in the interior of the simplex S_n exists.

Informations about the stability of the fixed points can be get from the corresponding eigenvalues.

For some values of the control u Table 3.2.2 shows the corresponding fixed points as well as the Jacobian and the eigenvalues for example **TEST#1**. It is easy to see that for increasing values of u the fixed point develops from an attractor (u = -4) to a saddle (u = 2.5) and then to a repellor (u = 6). For example **TEST#31** we have qualitatively the same results which can be seen from Table 3.2.3.

FIXED POINT FOR DIAGONAL CONTROL VALUE = -6.000000E+00

6.676395E-02	5.647574E-02	1.001094E-01	1.289675E-01
3.622765E-02	3.659248E-02	6.048887E-02	1.942722E-01
1.659613E-01	1.541408E-01		

EIGENVALUES OF THE REPLICATOR EQUATION FOR ABOVE FIXED POINT

REAL	IMAGINARY	INDICATOR
-1.652087E+00	0.000000E+00	1.000000E+00
-1.456159E+00	0.000000E+00	1.000000E+00
-8.036321E-01	0.000000E-00	1.000000E+00
-3.514265E-02	0.000000E+00	1.000000E+00
-6.301777E-01	0.000000E+00	1.000000E+00
-2.642226E-01	0.000000E+00	1.000000E+00
-3.480596E-01	0.000000E+00	1.000000E+00
-4.372603E-01	0.000000E+00	1.000000E+00
-3.953311E-01	0.000000E+00	1.000000E+00

FIXED POINT FOR DIAGONAL CONTROL VALUE = -4.100000E+00

2.182102E-03	3.411289E-02	3.547504E-02	8.398681E-02
1.887011E-01	1.917111E-01	6.150402E-02	1.532130E-01
1.228776E-01	1.262363E-01		

EIGENVALUES OF THE REPLICATOR EQUATION FOR ABOVE FIXED POINT

REAL	IMAGINARY	INDICATOR
-1.736925E+00	0.000000E+00	1.000000E+00
-9.948613E-01	0.000000E+00	1.000000E+00
-8.935437E-01	0.000000E+00	1.000000E+00
-7.764422E-03	0.000000E+00	1.000000E+00
1.743881E-01	0.000000E+00	1.000000E+00
-3.211513E-01	0.000000E+00	1.000000E+00
-1.591105E-01	0.000000E+00	1.000000E+00
-2.823447E-01	0.000000E+00	1.000000E+00
-9.768120E-02	0.000000E+00	1.000000E+00

FIXED POINT FOR DIAGONAL CONTROL VALUE = -4.800000E+00

1.427057E-02	6.154207E-02	2.850259E-02	6.532997E-02
2.674438E-01	2.678321E-01	6.534740E-02	9.298981E-02
6.692381E-02	6.981795E-02		

EIGENVALUES OF THE REPLICATOR EQUATION FOR ABOVE FIXED POINT

REAL	IMAGINARY	INDICATOR
-2.630174E+00	0.000000E+00	1.000000E+00
-7.006674E-01	0.000000E+00	1.000000E+00
-5.470944E-01	0.000000E+00	1.000000E+00
-3.935204E-01	0.000000E+00	1.000000E+00
6.116959E-01	0.000000E+00	1.000000E+00
-3.204006E-01	0.000000E+00	1.000000E+00
-5.877762E-02	0.000000E+00	1.000000E+00
-1.804502E-01	0.000000E+00	1.000000E+00
-1.470572E-01	0.000000E+00	1.000000E+00

FIXED POINT FOR DIAGONAL CONTROL VALUE = 1.300000E+00

2.416256E-01	2.137107E-01	2.787805E-03	3.840703E-03
3.612161E-02	6.051675E-02	3.039727E-01	4.268348E-02
2.985291E-02	6.488777E-02		

EIGENVALUES OF THE REPLICATOR EQUATION FOR ABOVE FIXED POINT

REAL	IMAGINARY	INDICATOR
3.022051E-01	0.000000E+00	1.000000E+00
5.282788E-01	0.000000E+00	1.000000E+00
-2.004848E-01	0.000000E+00	1.000000E+00
1.585316E-01	0.000000E+00	1.000000E+00
-1.061874E-01	0.000000E+00	1.000000E+00
1.088075E-01	0.000000E+00	1.000000E+00
-2.231675E-02	0.000000E+00	1.000000E+00
1.716702E-02	0.000000E+00	1.000000E+00
-4.925355E-02	0.000000E+00	1.000000E+00

FIXED POINT FOR DIAGONAL CONTROL VALUE = 1.600000E+00

2.434628E-01	1.860657E-01	2.181202E-03	1.563871E-02
4.863224E-02	7.144007E-02	2.686123E-01	7.394409E-02
4.110900E-02	5.394350E-02		

EIGENVALUES OF THE REPLICATOR EQUATION FOR ABOVE FIXED POINT

REAL	IMAGINARY	INDICATOR
5.651789E-01	0.000000E+00	1.000000E+00
-2.280404E-01	0.000000E+00	1.000000E+00
3.925237E-01	0.000000E+00	1.000000E+00
3.896886E-02	0.000000E+00	1.000000E+00
2.148837E-01	0.000000E+00	1.000000E+00
1.975495E-01	0.000000E+00	1.000000E+00
-1.120171E-02	0.000000E+00	1.000000E+00
-9.657784E-02	0.000000E+00	1.000000E+00
-9.436236E-02	0.000000E+00	1.000000E+00

FIXED POINT FOR DIAGONAL VALUE = 7.000000E+00

1.151299E-01	1.118817E-01	1.080621E-01	1.117441E-01
9.867404E-02	8.357283E-02	1.420841E-01	3.547664E-02
8.654879E-02	1.068256E-01		

EIGENVALUES OF THE REPLICATOR EQUATION FOR ABOVE FIXED POINT

REAL	IMAGINARY	INDICATOR
1.073324E+00	0.000000E+00	1.000000E+00
1.117693E+00	0.000000E+00	1.000000E+00
4.579958E-01	0.000000E+00	1.000000E+00
3.528669E-01	0.000000E+00	1.000000E+00
6.923334E-01	0.000000E+00	1.000000E+00
1.401888E-01	0.000000E+00	1.000000E+00
8.550338E-01	0.000000E+00	1.000000E+00
1.701911E-01	0.000000E+00	1.000000E+00
9.199096E-01	0.000000E+00	1.000000E+00

Table 3.2.3 Fixed points and eigenvalues of example **TEST#31** for different values of u

3.3. Cooperation and Competition

Within this section we want to clear the question what happens to the set of fixed points, and especially to the set of stable fixed points, by applying a homogeneous replication rate control.

Doing this we follow the point of view, that a stable fixed point represents those replicator types within the total population with their proportions which behave cooperative. For this analysis we first consider for a homogeneous replication rate control the two extreme cases of very large and very small u.

In this case we have a fixed point at the barycentre of S_n. The division of the denominator and nominator of the fixed point relation

$$p=\frac{\left(\sum_i B_i u^{n-1-i}\right)e}{\left(e,\left(\sum_i B_i u^{n-1-i}\right)e\right)}$$

by u^{n-1} yields

$$p=\frac{\left(\sum_i B_i u^{-i}\right)e}{\left(e,\left(\sum_i B_i u^{-i}\right)e\right)}$$

and therefore because of $B_0 = E$

$$\lim_{u\to+\infty} p=\frac{e}{(e,e)}$$

or

$$\lim_{u\to+\infty} p_i=\frac{1}{n} \quad , \quad i=1,\ldots,n \quad .$$

The stability of this fixed point p in int(S_n) is dependent on u.

(i) For u very small p is an attractor. p is even an ESS. Therefore int(S_n) is the basin of attraction of p. The vertices of the simplex are repellors.

(ii) For u very large we have the opposite constellation. p is a repellor. The vertices of the simplex are attractors.

These constellations are depicted in Figure 3.3.1 for $n = 3$. We consider the set of stable fixed points on S_n as a global state dependent on u. Then we define by (i) a global state of uniform cooperation, and by (ii) a global state of uniform competition.

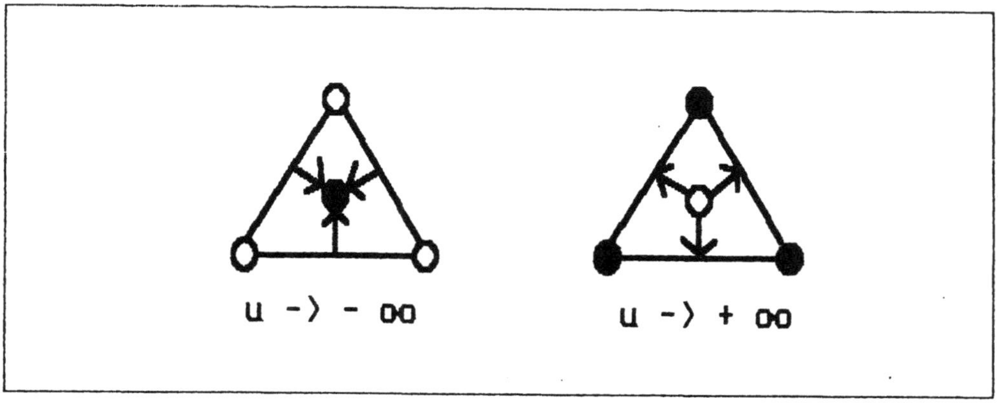

Figure 3.3.1 Fixed point distribution for homogeneous replication rate control, u very small (cooperation), u very large (competition)

This feature can be shown very easily e.g. for non-negative symmetric interaction matrices.

First we consider the case $u > 0$. With the fitness function

$$f_i = (Gx)_i + ux_i = u\left[\frac{(GX)_i}{u} + x_i\right]$$

we have the replicator equation

$$\dot{x}_i = x_i u\left[\frac{(Gx)_i}{u} + x_i - \frac{(x,Gx)}{u} - \sum_j x_j^2\right].$$

The factor u performs a time transformation and can be omitted. Then the fitness function of the total gene pool is given by

$$\Phi = \frac{(x,Gx)}{u} + \sum_j x_j^2.$$

For large u we have

$$\lim_{u\to\infty} \Phi = \sum_j x_j^2.$$

Because of the fundamental theorem

$$\Phi \to \max$$

this yields for the stable fixed point

$$x_i = 1 \quad , \quad x_j = 0 \quad , \quad j = 1,\ldots,n \text{ and } j \neq i \quad \forall i \quad \text{for } u \to \infty \quad .$$

The case $u < 0$ can be treated in a similar way. The fitness function

$$f_i = (Gx)_i - |u| x_i = |u| \left[\frac{(Gx)_i}{|u|} - x_i \right]$$

yields the replicator equation

$$\dot{x}_i = x_i |u| \left[\frac{(Gx)_i}{|u|} - x_i - \frac{(x, Gx)}{|u|} + \sum_j x_j^2 \right].$$

Regarding $|u|$ again as a time transformation we have the fitness function of the total gene pool

$$\Phi = \frac{(x, Gx)}{|u|} - \sum_j x_j^2$$

and therefore for very large $|u|$

$$\lim_{|u| \to \infty} = - \sum_j x_j^2 .$$

Because of

$$\Phi \to \max$$

This yields for the stable fixed point

$$x_i = \frac{1}{n} \quad , \quad i = 1,\ldots,n \quad \text{for } |u| \to \infty \quad .$$

Thus, dependent on the control u one gets a scaling of the global state shown in Figure 3.3.2.

increasing competition
$u \rightarrow +\infty$

$-\infty \leftarrow u$
increasing cooperation

Figure 3.3.2 Scaling of the global state via replication rate control

For very large values of the homogeneous replication rate control u only one replicator type survives dependent on the initial state. Therefore we use for the control u also the notion " **selection pressure** ".

For symmetric real interaction matrices G and homogeneous replication rate control u we observed the following fact:

For monotone increasing replication rate control u the number of stable fixed points on the simplex S_n is monotone increasing. Conversely, the number of stable fixed points on the simplex S_n decreases monotonously with monotone decreasing replication rate control.

For the example **TEST#31** Figure 3.3.3 shows the influence of the change of the selection pressure on the number and the position of the stable fixed points. The replicator types (vertices) belonging to a stable fixed point are characterized by a cluster (encircled by heavy lines). By this figure the above mentioned observation will be visually emphasized.

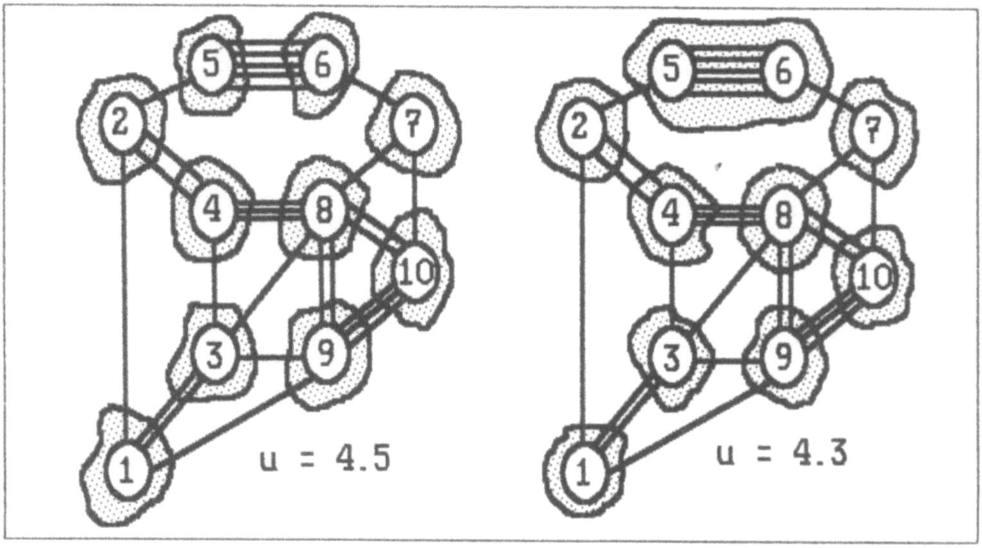

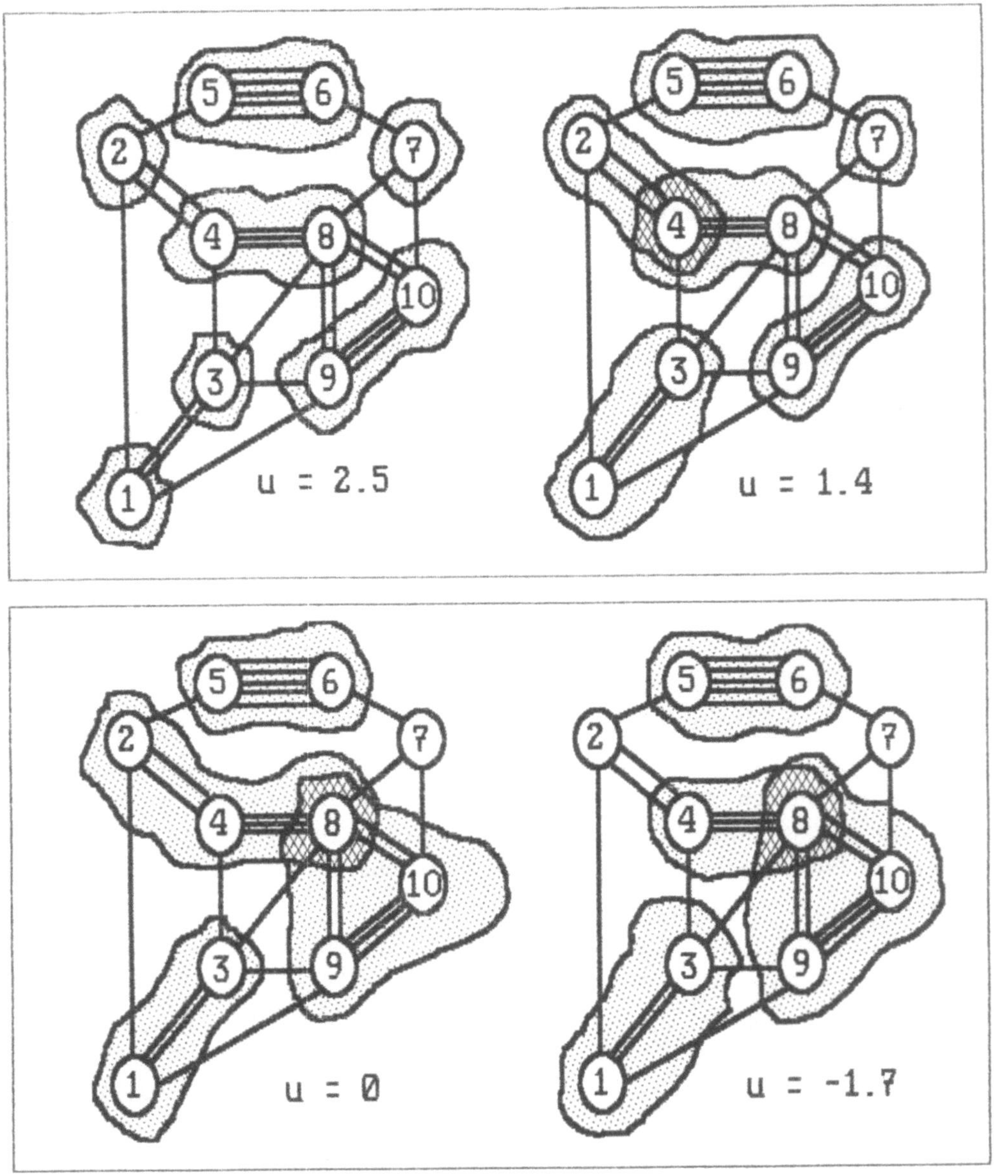
u = 2.5
u = 1.4
u = 0
u = -1.7

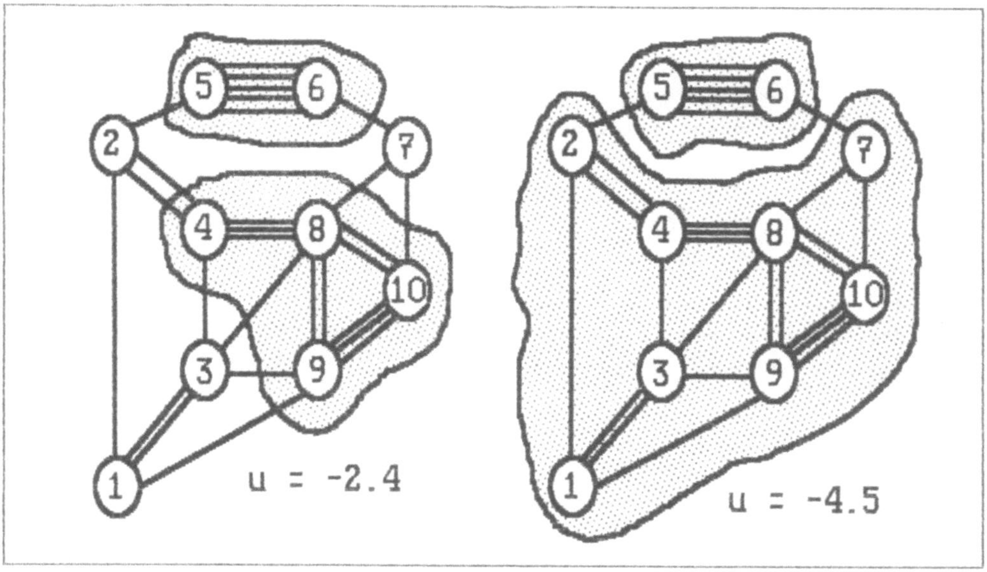

Figure 3.3.3 Stable fixed points (clusters) dependent on the selection pressure u

3.4. Boundary Bifurcations

How to explain now the change in the number of stable fixed points with changing selection pressure?

We want to clear this question by means of a small example. Let us consider the interaction matrix G with homogeneous replication rate control u

$$G = \begin{pmatrix} u & a \\ b & u \end{pmatrix}.$$

A fixed point is given by

$$p_1 = \frac{u-a}{2u-a-b}, \quad p_2 = \frac{u-b}{2u-a-b}.$$

We consider this fixed point for all $p_1 + p_2 = 1$, i.e. without the restriction that p is in S_2. Without loss of generality we assume $a > b$. By means of the transformation

$$y_1 = p_1 - \frac{1}{2} \;, \quad y_2 = -y_1 \;, \quad x = u - \frac{a+b}{2}$$

we have the fixed point relations in the form

$$y_1 = \frac{1}{2} + \frac{b-a}{4} \cdot \frac{1}{x} \;, \quad y_2 = \frac{1}{2} + \frac{a-b}{4} \cdot \frac{1}{x} \;.$$

These are hyperbola equations. The graph of the fixed points including their stability (repellor, and attractor respectively) versus the selection pressure is shown in Figure 3.4.1.

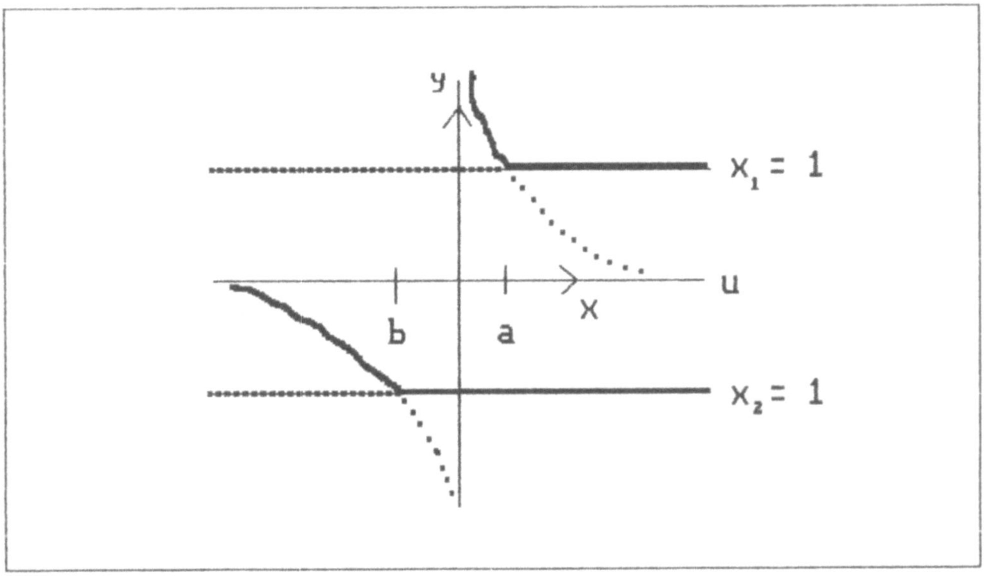

Figure 3.4.1 Fixed points versus selection pressure (repellor: dotted line, attractor: solid line)

If we show the flow of the replicator equation for this example on a real straight line we have the different qualitative constellations depicted in Figure 3.4.2.

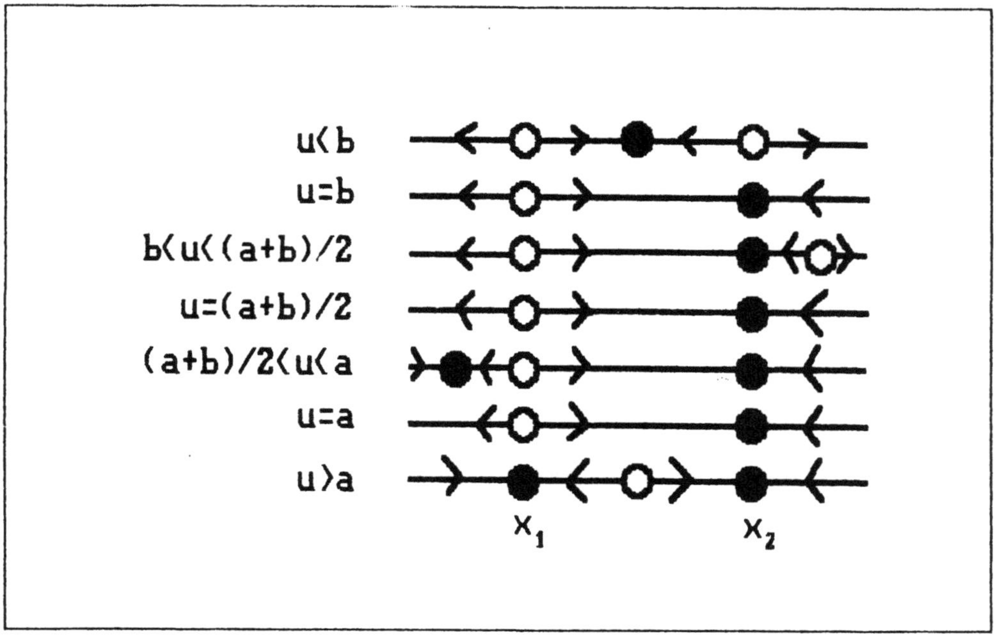

Figure 3.4.2 Fixed point distributions for different values of the selection pressure

It is quite obvious that on the boundary of the simplex S_n a stability exchange between fixed points occurs. For the given example it happens for $u = a$ and for $u = b$.

For $a = 2$ and $b = 1$, i.e.

$$G = \begin{pmatrix} 2 & u \\ u & 1 \end{pmatrix}$$

fixed points for different values of u in the range (-100, 100) are presented in Table 3.4.1. The graph of the fixed point components p_1, p_2 is shown for $0 < u < 3$ in Figure 3.4.3.

```
FIXED POINT FOR DIAGONAL CONTROL VALUE = -100

 5.024631E-01  4.975369E-01

FIXED POINT FOR DIAGONAL CONTROL VALUE = 0

 6.666667E-01  3.333333E-01

FIXED POINT FOR DIAGONAL CONTROL VALUE = 1

 1.000000e+00  0.000000e+00

FIXED POINT FOR DIAGONAL CONTROL VALUE = 1.4

-9.999999E-99  -9.999999E-99

FIXED POINT FOR DIAGONAL CONTROL VALUE = 1.6

-9.999999E-99  -9.999999E-99

FIXED POINT FOR DIAGONAL CONTROL VALUE = 2

 0.000000E+00  1.000000E+00

FIXED POINT FOR DIAGONAL CONTROL VALUE = 3

 3.333333E-01  6.666667E-01

FIXED POINT FOR DIAGONAL CONTROL VALUE = 100

 4.974619E-01  5.025381E-01
```

Table 3.4.1 Fixed points for different values of the selection pressure

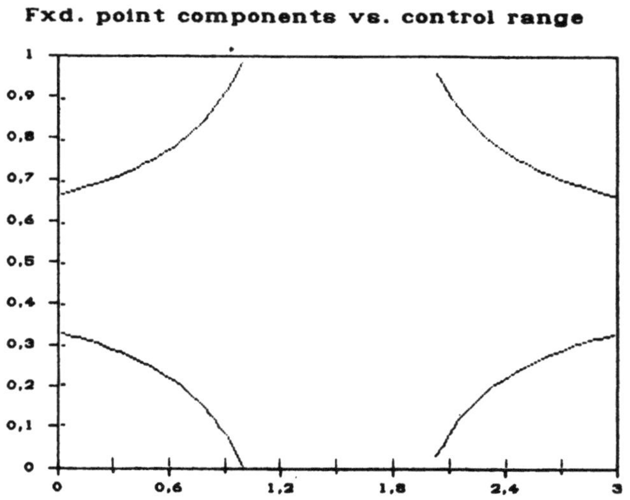

Figure 3.4.3 Graph of fixed point for $0 < u < 3$

Considering for this example the graph of the fixed point on the simplex S_n it is possible to show a bifurcation diagram corresponding to Figure 3.4.4.

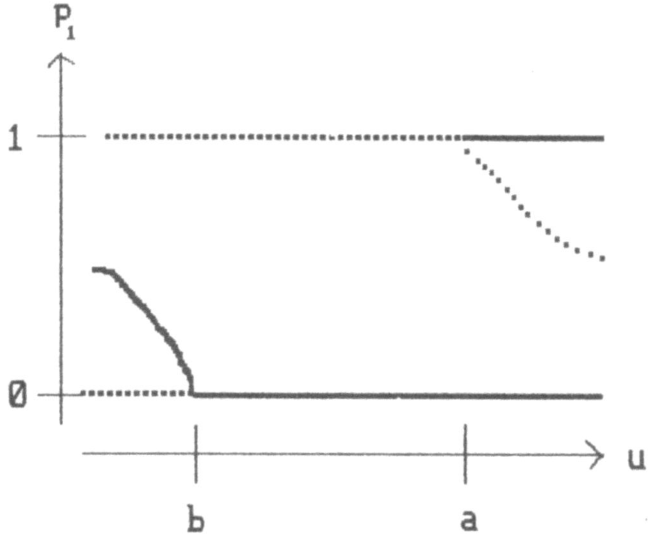

Figure 3.4.4 Bifurcation diagram (stable: solid line, unstable: dotted line)

Let us return once more to the observed fact of the monotony of the number of stable fixed points dependent on the selection pressure from Section 3.3. To underline this observation we present schematically the number of stable fixed points versus the control variable u. For the above given example and for the example **TEST#1** we get the graphs shown in figs. 3.4.5. and 3.4.6. The encircled numbers represent the replicator types belonging to a stable fixed point.

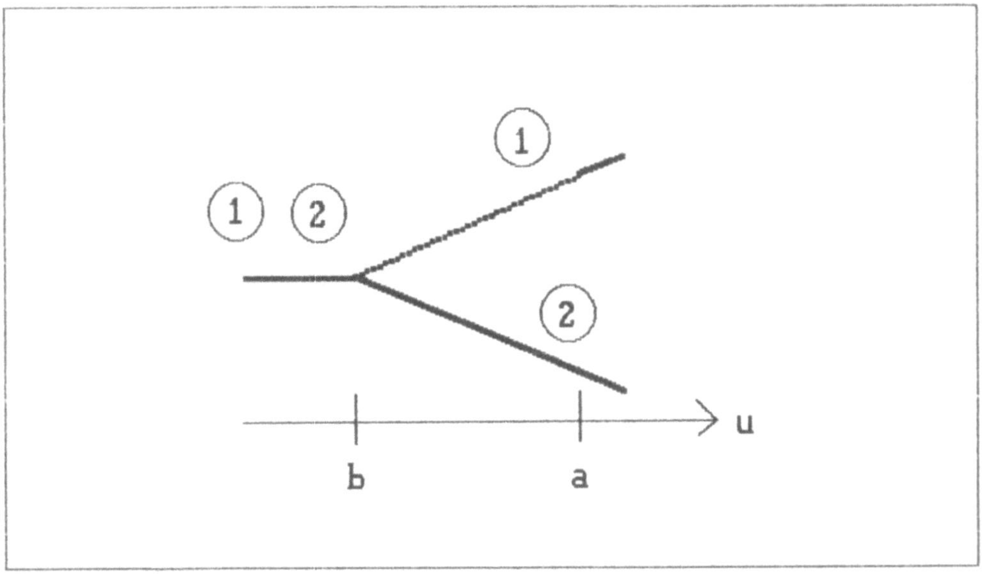

Figure 3.4.5 Bifurcation tree (stable: solid line, unstable: dotted line)

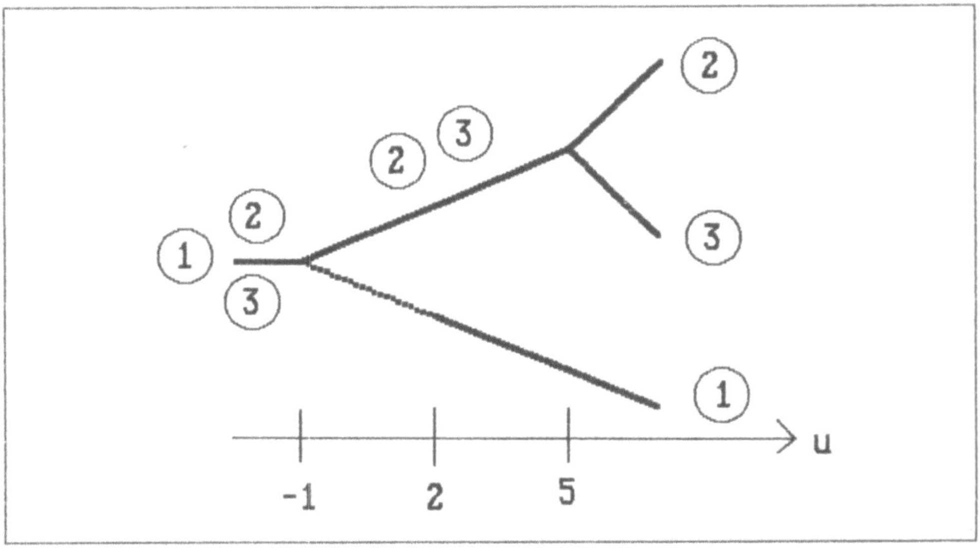

Figure 3.4.6 Bifurcation tree for TEST#1 (stable: solid line, unstable: dotted line)

3.5. Hopf Bifurcations

Hopf bifurcations play an important role with respect to limit cycles or periodic orbits under the influence of variations or control of parameters of the replicator equation.

Within this section we do not dig to much into details but show the application and effectiveness of the used programs by means of the example **HYPER1**. A detailed treatment of the Hopf bifurcation is given e.g. by Marsden & McCracken (1976).

To give a clear idea a Hopf bifurcation occurs under the following circumstances:

Let us assume that a system rests in a stable fixed point. All eigenvalues of this fixed point have negative real parts. Some eigenvalues should be conjugate complex. If one or more parameters of the system will be changed so that one complex conjugate eigenvalue traverses from the left-hand side to the right-hand side of the complex plane and the remaining eigenvalues will stay in the left-hand side, then a Hopf bifurcation occurs.

This means that a stable or unstable limit cycle arises. The stability of the limit cycle has to be checked separately.

These facts hold only for the linearized system at the fixed point. Global features can be extracted by an analysis of the global stability (Jansen (1982,1987), Peschel & Mende (1983)) or permanence (Amann & Hofbauer (1985)).

We want to clarify the steps for the computation of a Hopf bifurcation by means of the hypercycle example **HYPER1**. First we generate an overview of the position of the fixed point in the interior of the simplex S_4. This is done by computing the polynomial coefficients of a fixed point of the controlled replicator equation as well as their zeros and the corresponding fixed points. This is shown in Table 3.5.1. Very helpful for this purpose is a graphical representation of the distribution of the fixed point components versus the control range as depicted in Figure 3.5.1. Clearly for **HYPER1** exists a fixed point at the barycentre of the simplex over the whole control range.

By means of a step by step eigenvalue analysis for different control values the bifurcation point $u_0 = 0$ can be computed. In this case the real eigenvalue is negative and the complex conjugate eigenvalue has a real part of zero up to rounding errors. Some values are presented in Table 3.5.2. It is interesting that the real parts of the eigenvalues depend linearly on the control variable u (see Section 3.1.).

The stability can be checked e.g. via a simulation. Figures 3.5.2 to 3.5.4 very clearly show for different control values u the limit cycles. The corresponding simulation parameters are mentioned.

The limit cycle runs for $u > 1/3$ to one of the vertices of the simplex (Amann & Hofbauer (1985)).

Here we want to point out also that for the Lotka-Volterra equation strange attractor may arise. Such a chaotic attractor was found by Arneodo, Coullet & Tresser (1980).

COEFFICIENTS OF COMPONENT POLYNOMIALS

1.000000E+00	-1.000000E+00	1.000000E+00	-1.000000E+00
1.000000E+00	-1.000000E+00	1.000000E+00	-1.000000E+00
1.000000E+00	-1.000000E+00	1.000000E+00	-1.000000E+00
1.000000E+00	-1.000000E+00	1.000000E+00	-1.000000E+00

REAL ZEROS OF COMPONENT POLYNOMIALS

1.000000E+00	-9.999999E-99	-9.999999E-99
1.000000E+00	-9.999999E-99	-9.999999E-99
1.000000E+00	-9.999999E-99	-9.999999E-99
1.000000E+00	-9.999999E-99	-9.999999E-99

FIXED POINT(1.00000000004)

2.500000E-01	2.500000E-01	2.500000E-01	2.500000E-01

Table 3.5.1 Fixed point components and their real zeros for **HYPER1**

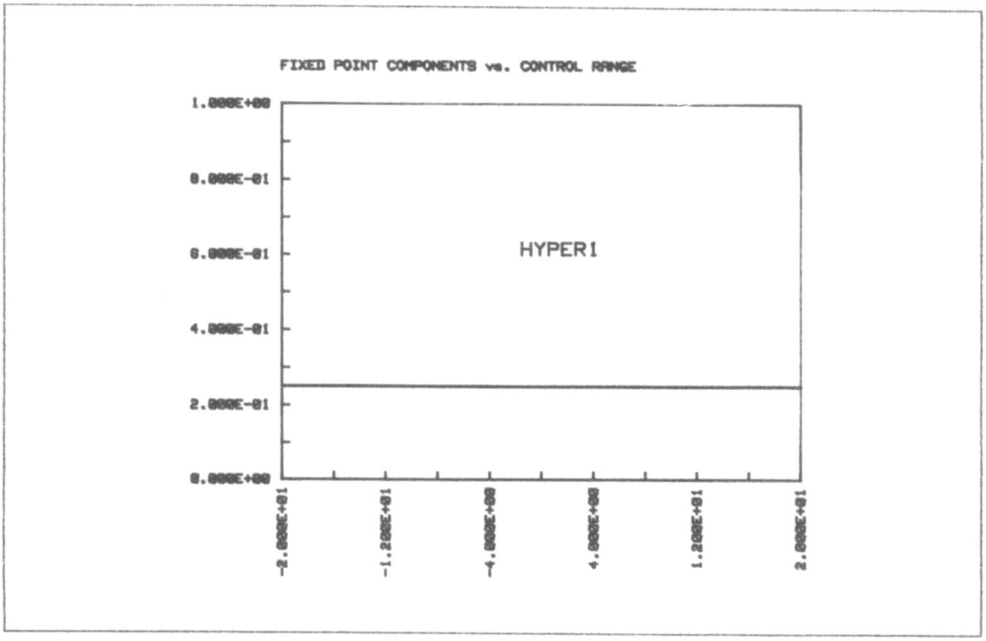

Figure 3.5.1 Fixed point for **HYPER1**

```
FIXED POINT FOR DIAGONAL CONTROL VALUE = -1.000000+00

 2.500000E-01  2.500000E-01  2.500000E-01  2.500000E-01

EIGENVALUES OF THE REPLICATOR EQUATION FOR ABOVE FIXED POINT
     REAL         IMAGINARY      INDICATOR

-5.000000E-01  0.000000E+00  1.000000E+00
-2.500000E-01  2.500000E-01  1.000000E+00
-2.500000E-01 -2.500000E-01  1.000000E+00
```

```
FIXED POINT FOR DIAGONAL CONTROL VALUE = 0.000000E+00

 2.500000E-01  2.500000E-01  2.500000E-01  2.500000E-01

EIGENVALUES OF THE REPLICATOR EQUATION FOR ABOVE FIXED POINT
     REAL         IMAGINARY      INDICATOR

-2.500000E-01  0.000000E+00  1.000000E+00
-5.542006E-13  2.500000E-01  1.000000E+00
-5.542006E-01  2.500000E-01  1.000000E+00
```

```
FIXED POINT FOR DIAGONAL CONTROL VALUE = 3.000000E-01

 2.500000E-01  2.500000E-01  2.500000E-01  2.500000E-01

EIGENVALUES OF THE REPLICATOR EQUATION FOR ABOVE FIXED POINT
     REAL         IMAGINARY      INDICATOR

-1.750000E-01  0.000000E+00  1.000000E+00
 7.500000E-02  2.500000E-01  1.000000E+00
 7.500000E-02  2.500000E-01  1.000000E+00
```

```
FIXED POINT FOR DIAGONAL CONTROL VALUE = 5.000000E-01

 2.500000E-01  2.500000E-01  2.500000E-01  2.500000E-01

EIGENVALUES OF THE REPLICATOR EQUATION FOR ABOVE FIXED POINT
     REAL         IMAGINARY      INDICATOR

-1.250000E-01  0.000000E+00  1.000000E+00
 1.250000E-01  2.500000E-01  1.000000E+00
 1.250000E-01  2.500000E-01  1.000000E+00
```

```
FIXED POINT FOR DIAGONAL CONTROL VALUE = 9.900000E-01

 2.500000E-01  2.500000E-01  2.500000E-01  2.500000E-01

EIGENVALUES OF THE REPLICATOR EQUATION FOR ABOVE FIXED POINT
     REAL          IMAGINARY      INDICATOR

-2.500000E-03  0.000000E+00  1.000000E+00
 2.475000E-01  2.500000E-01  1.000000E+00
 2.475000E-01  2.500000E-01  1.000000E+00
```

```
FIXED POINT FOR DIAGONAL CONTROL VALUE = 1.100000E+00

 2.500000E-01  2.500000E-01  2.500000E-01  2.500000E-01

EIGENVALUES OF THE REPLICATOR EQUATION FOR ABOVE FIXED POINT
     REAL          IMAGINARY      INDICATOR

 2.500000E-02  0.000000E+00  1.000000E+00
 2.750000E-01  2.500000E-01  1.000000E+00
 2.750000E-01  2.500000E-01  1.000000E+00
```

Table 3.5.2 Eigenvalues versus selection pressure for **HYPER1**

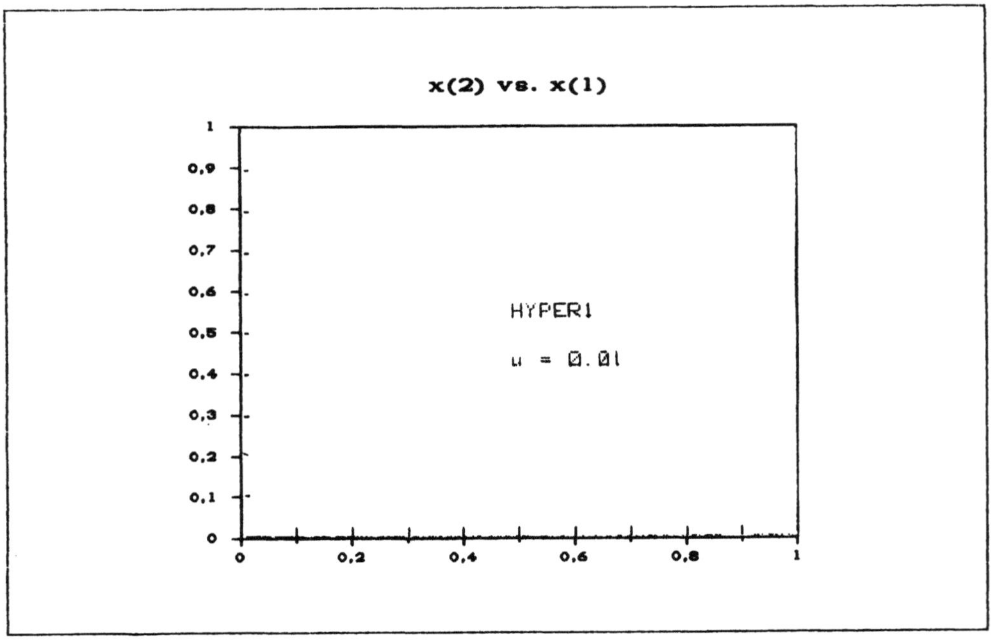
x(2) vs. x(1)
1
0,9
0,8
0,7
0,6
0,5
0,4
0,3
0,2
0,1
0
0
0,2
0,4
0,6
0,8
1
HYPER1
u = 0.01

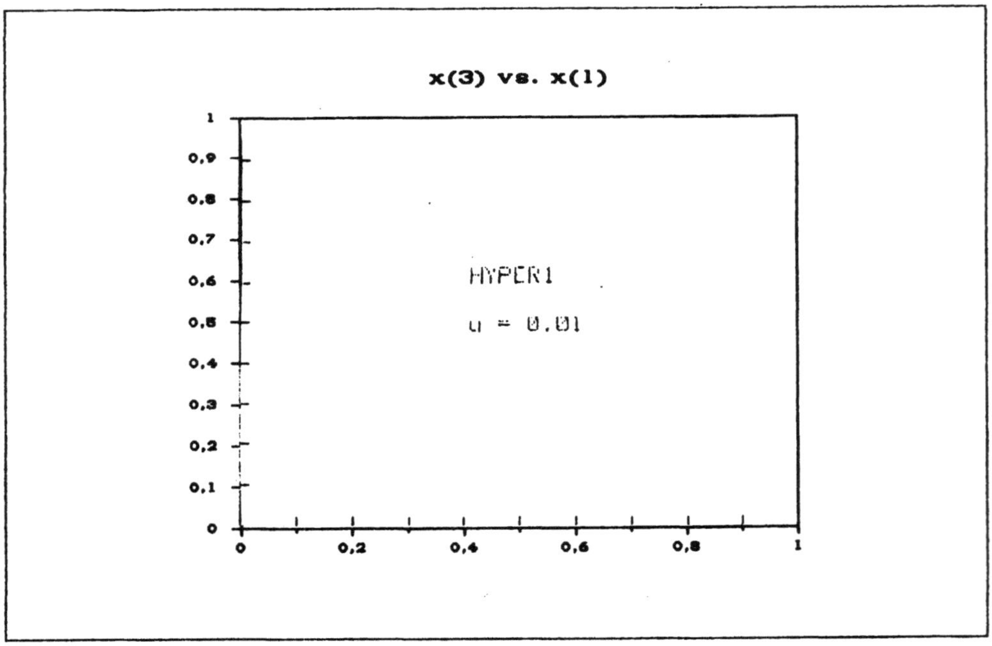
x(3) vs. x(1)
1
0,9
0,8
0,7
0,6
0,5
0,4
0,3
0,2
0,1
0
0
0,2
0,4
0,6
0,8
1
HYPER1
u = 0.01

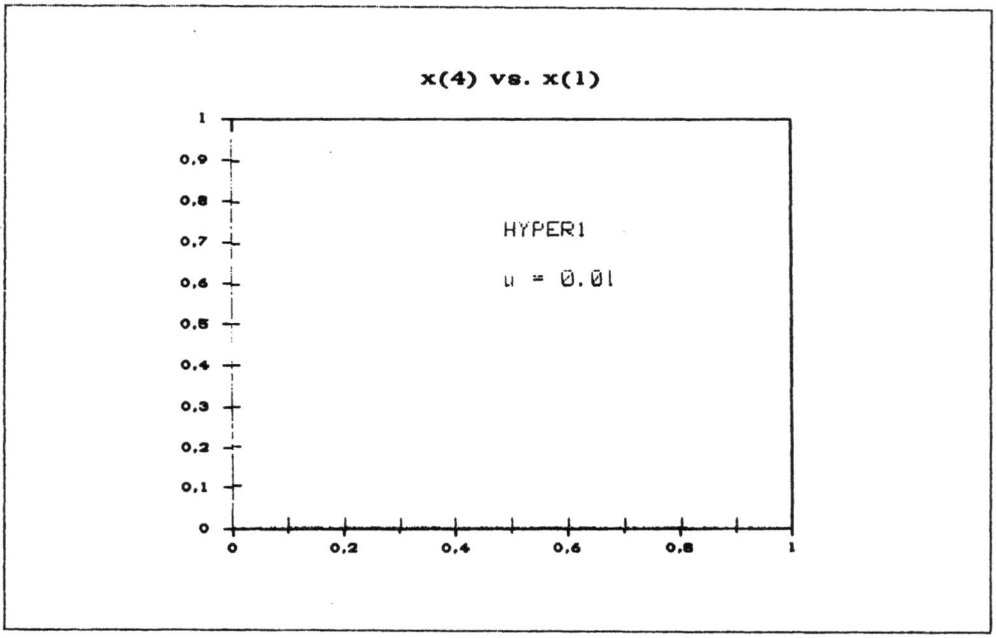

SAMPLE IDENTIFIER: HYPER1
NUMBER OF OBJECTS = 4
LOWER BOUND FOR INTEGRATION = 0
UPPER BOUND FOR INTEGRATION = 400
STEP SIZE FOR INTEGRATION = .25
MAX # OF INTEGRATION STEPS = 2000
PRECISION OF RESULTS FOR INTEGRATION = .0000001
CONTROL PARAMETER = .01

Figure 3.5.2 Limit cycle for u = 0.01

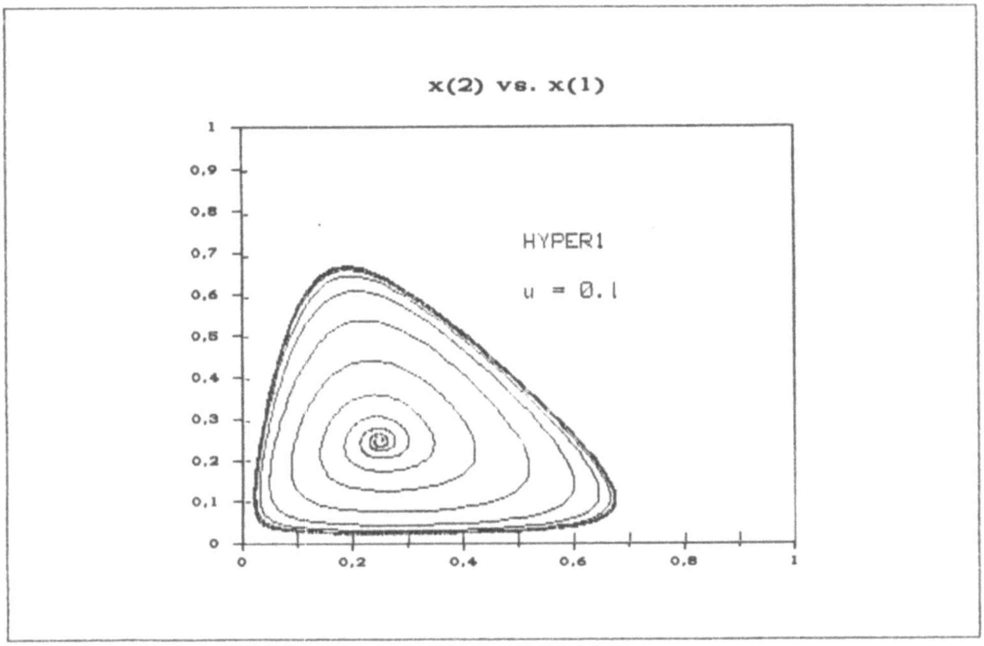
x(2) vs. x(1)
HYPER1
u = 0.1
1
0,9
0,8
0,7
0,6
0,5
0,4
0,3
0,2
0,1
0
0
0,2
0,4
0,6
0,8
1

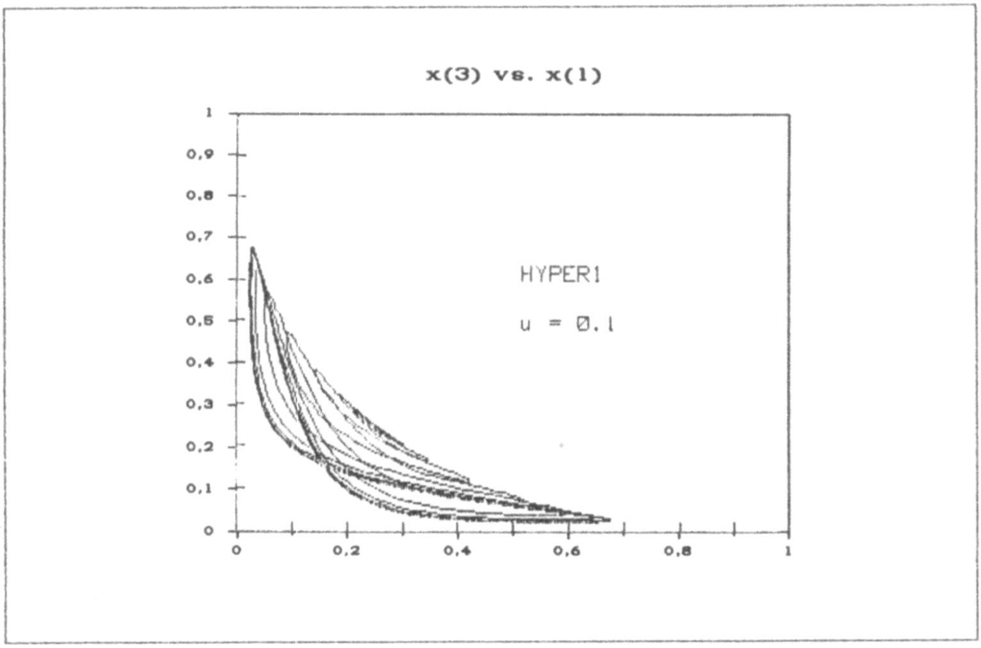
x(3) vs. x(1)
HYPER1
u = 0.1
1
0,9
0,8
0,7
0,6
0,5
0,4
0,3
0,2
0,1
0
0
0,2
0,4
0,6
0,8
1

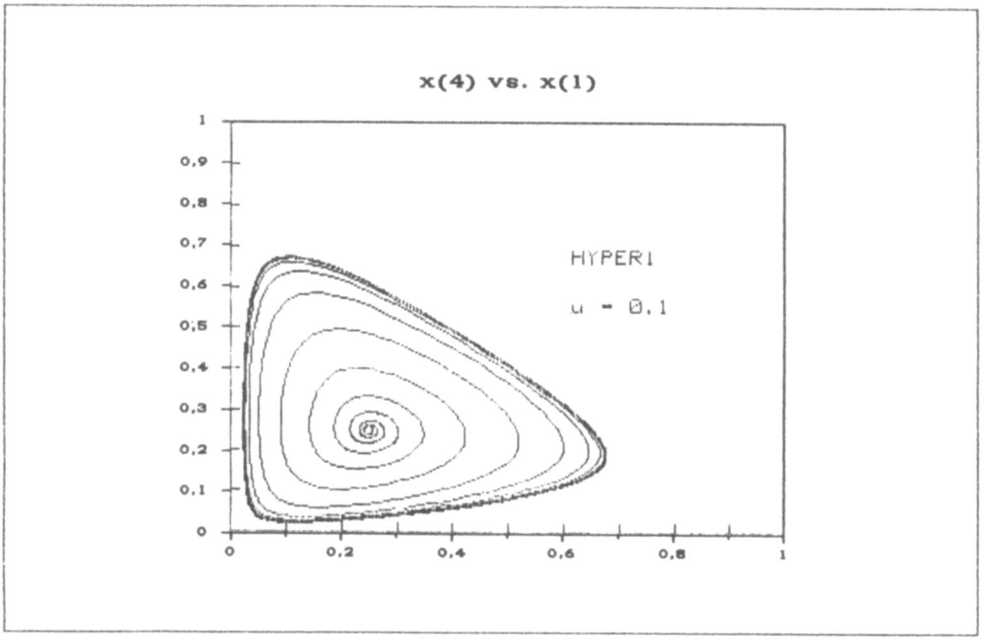


```
SAMPLE IDENTIFIER: HYPER1
NUMBER OF OBJECTS =  4
LOWER BOUND FOR INTEGRATION =   0
UPPER BOUND FOR INTEGRATION =   400
STEP SIZE FOR INTEGRATION =   .25
MAX # OF INTEGRATION STEPS =  2000
PRECISION OF RESULTS FOR INTEGRATION =  .0000001
CONTROL PARAMETER =  .1
```

Figure 3.5.3 Limit cycle for $u = 0.1$

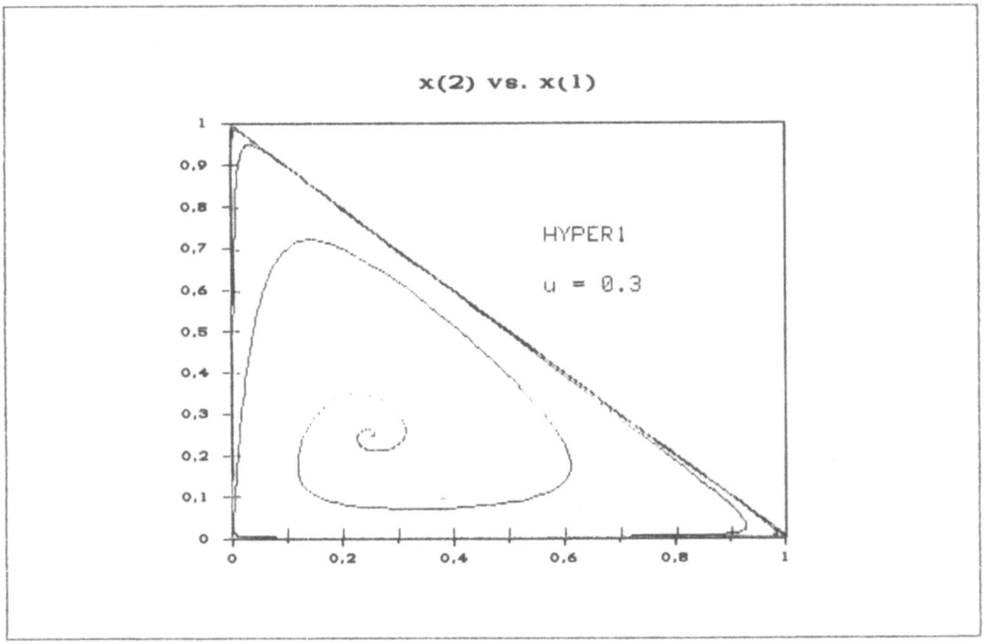
x(2) vs. x(1)
HYPER1
u = 0.3
1
0,9
0,8
0,7
0,6
0,5
0,4
0,3
0,2
0,1
0
0
0,2
0,4
0,6
0,8
1

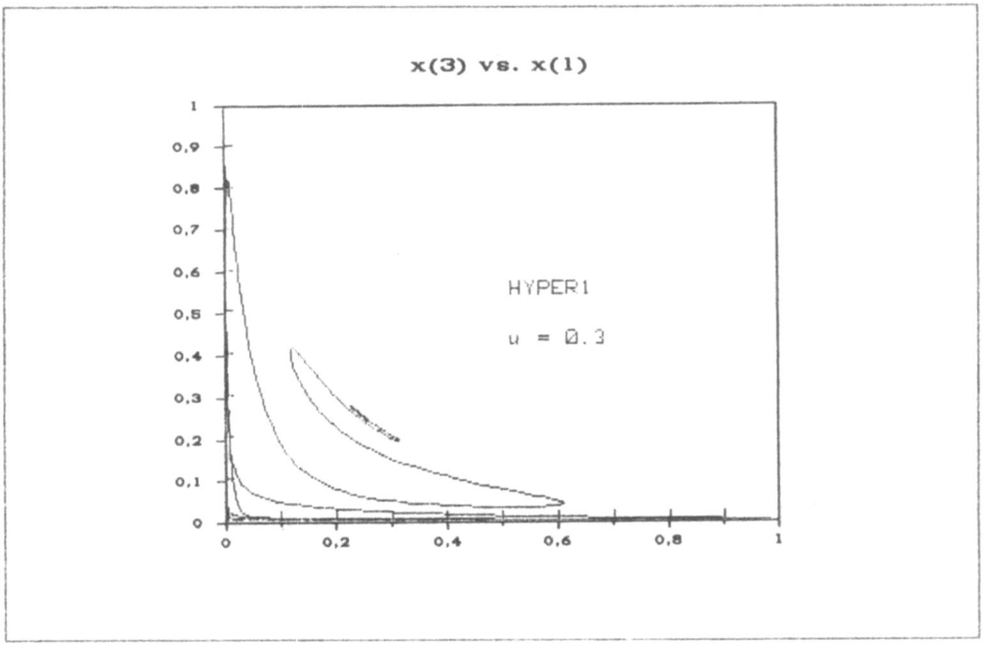
x(3) vs. x(1)
HYPER1
u = 0.3
1
0,9
0,8
0,7
0,6
0,5
0,4
0,3
0,2
0,1
0
0
0,2
0,4
0,6
0,8
1

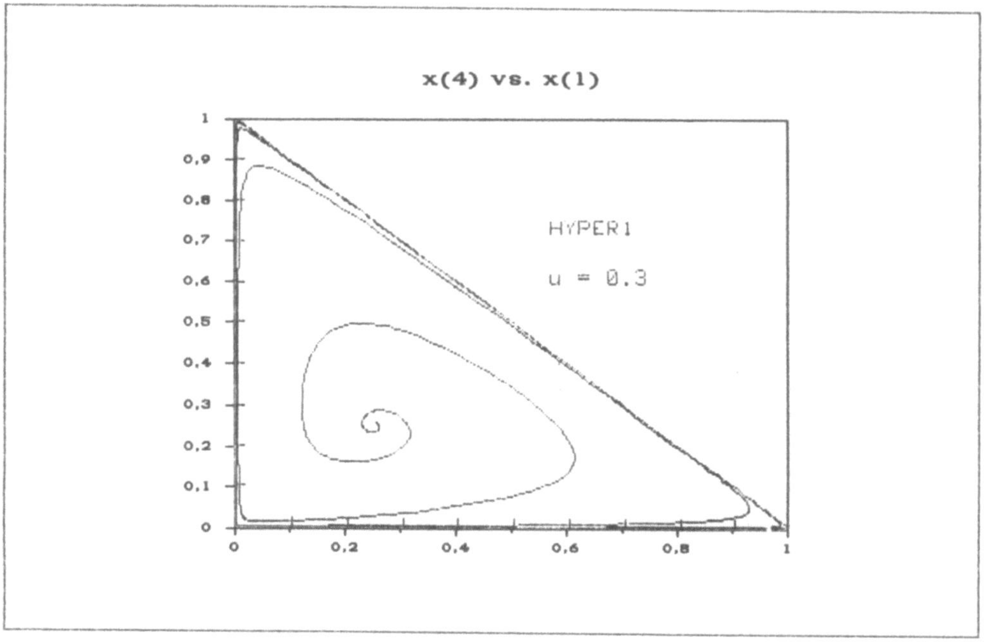

SAMPLE IDENTIFIER: HYPER1
NUMBER OF OBJECTS = 4
LOWER BOUND FOR INTEGRATION = 0
UPPER BOUND FOR INTEGRATION = 400
STEP SIZE FOR INTEGRATION = .25
MAX # OF INTEGRATION STEPS = 2000
PRECISION OF RESULTS FOR INTEGRATION = .0000001
CONTROL PARAMETER = .3

Figure 3.5.4 Limit cycle for u = 0.3

Connected to Hopf bifurcations and limit cycles it is interesting to construct such controls for arbitrary interaction matrices that limit cycles arise for suitable values of the controls.

This is possible in any case if we allow that the control variables u_i can influence any interaction coefficient.

The basic idea consists in impressing upon a given replicator equation a hypercycle structure (Eigen & Schuster (1979)) by the control variables.

Because of the invariance property (see Section 2.7.) to generate zero elements of the interaction matrix in the sense of row permutations such that e.g.

$$G=\begin{pmatrix} * & 0 & * & \cdots & * & * \\ * & * & 0 & \cdots & * & * \\ . & . & . & \cdots & . & . \\ * & * & * & \cdots & * & 0 \\ 0 & * & * & \cdots & * & * \end{pmatrix}.$$

The * indicates arbitrary elements. If we insert control variables u_i at the zero element positions we have the interaction matrix

$$G=\begin{pmatrix} * & u_1 & * & \cdots & * & * \\ * & * & u_2 & \cdots & * & * \\ . & . & . & \cdots & . & . \\ * & * & * & \cdots & * & u_{n-1} \\ u_n & * & * & \cdots & * & * \end{pmatrix}.$$

This type of control we denote as a hypercycle control. As in the case of the self-replication rate control we want to analyze the two extreme cases for very large and very small control values u_i, i.e. we consider

$$u_i \to \pm\infty \quad .$$

Then we have

$$u = u_i \to \pm\infty \quad .$$

The replicator equation, controlled in such a way, has with the fitness function f_i

$$f_i = (Gx)_i + sign(u)|u|x_{i+1}, x_{n+1} = x_1$$

or

$$f_i = |u|\left[\frac{(Gx)_i}{|u|} + sign(u)x_{i+1}\right]$$

the form

$$\dot{x}_i = x_i|u|\left[(Gx)_i + sign(u)x_{i+1} - \sum_i x_i\frac{(Gx)_i}{|u|} - sign(u)\sum_i x_i x_{i+1}\right].$$

The factor |u| performs a time transformation, and can be omitted. For very large and very small u this yields the replicator equation

$$\dot{x}_i = x_i\left(x_{i+1} - \sum_i x_i x_{i+1}\right).$$

But this is a hypercycle corresponding to Figure 3.5.5.

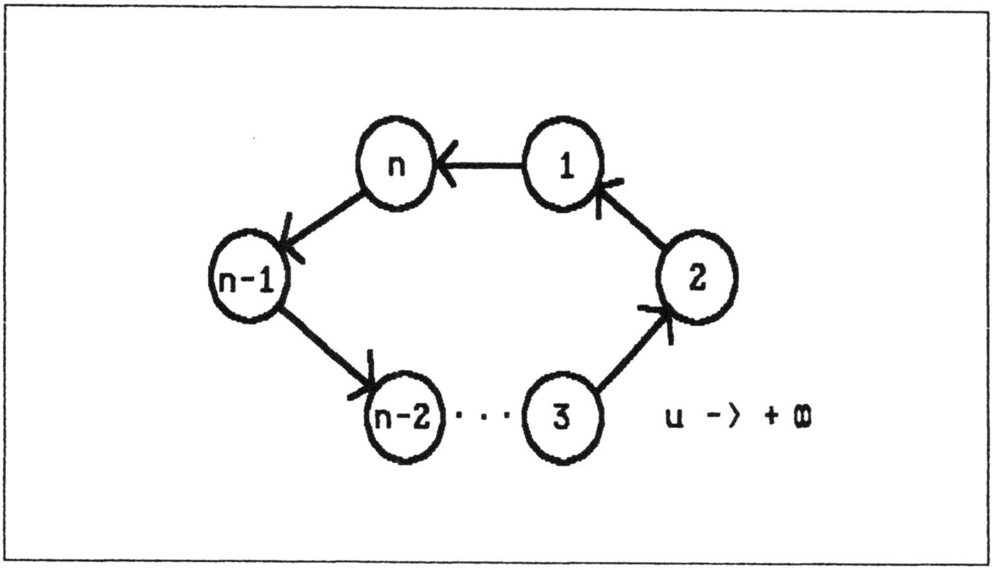

Figure 3.5.5 Hypercycle for very large u

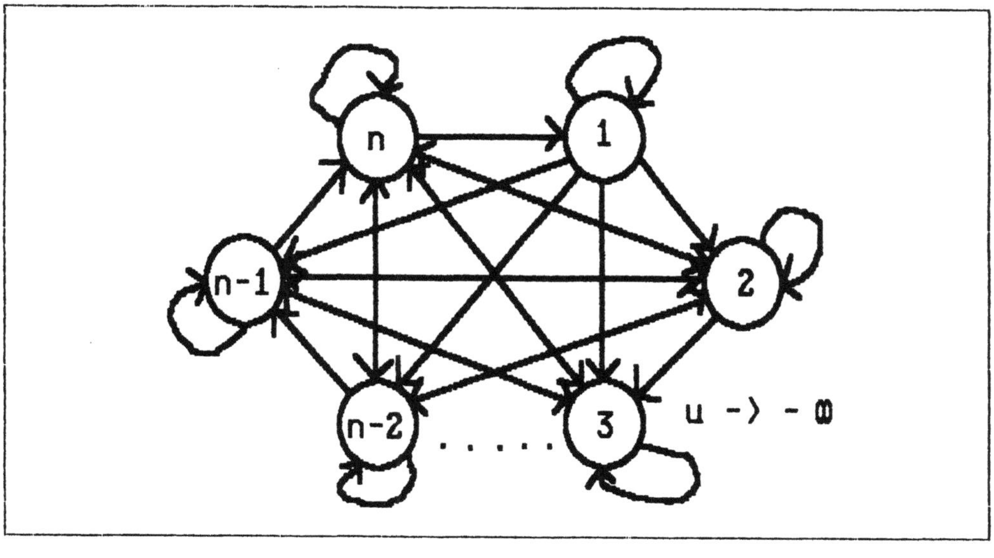

Figure 3.5.6 Complement of the hypercycle for very small u

For the case of very small u we have the replicator equation

$$\dot{x}_i = x_i\left(-x_{i+1} + \sum_i x_i x_{i+1}\right).$$

Adding to every column of the interaction matrix a constant $a_j = 1$ yields the graph depicted in Figure 3.5.6. This is the complement of the hypercycle graph of Figure 3.5.5 to a complete graph.

For very large u we get for $n > 4$ a stable limit cycle. For n = 2,3,4 all orbits run to the barycentre of the simplex. Therefore, no replicator type will be extincted. The replicator types behave cooperative.

For very small u we have the following feature. The orbits run from the barycentre to the boundary of the simplex. Whole faces of the simplex are fixed. These fixed faces consist of the partial interaction matrices with all elements $G_{ij} = 1$. For $n = 4$ e.g. the edges (1,3) and (2,4) are fixed.

4. The Diversified Replicator Equation

Up to now we assumed that the evolution of the single replicator types proceeds within a single "area". The term single "area" may be thought of ecologically closed regions, well stirred bioreactors with a corresponding dilution flow (constant organization), or a gene locus.

But what happens if we assume that evolution takes place within a number of "areas" which are connected to each other ?

If we follow this idea it is quite natural to use for the description of the evolution process a replicator equation for each of these "areas" with

$$\dot{x}_{ik} = x_{ik}(f_{ik} - \Phi_i) \text{ or } x'_{ik} = x_{ik}\frac{f_{ik}}{\Phi_i}.$$

This equation we denote as the diversified replicator equation.

The subscript $i = 1,\dots,n$ denotes the replicator type, and the subscript $k = 1,\dots,m$ the "area".

As constant organization we understand the condition

$$\sum_k x_{ik} = 1.$$

This means that every replicator type is defined on a unit simplex

$$S^i_m = \left\{(x_{i1},\dots,x_{im}) : x_{ik} \geq 0, \sum_k x_{ik} = 1\right\}.$$

The whole state space S of the diversified replicator equation is the direct sum of the disjoint simplices

$$S = \bigoplus_i S^i_m.$$

f_{ik} is the mean fitness of the replicator type i in "area" k. The total mean fitness of the replicator type i is given by

$$\Phi_i = \sum_k x_{ik} f_{ik}.$$

For the total fitness of the population, and the gene pool respectively, we have

$$\Phi = \sum_i \Phi_i .$$

Fixed points p_{ik} of the diversified replicator equation are characterized by the relations

$$p_{ik} = 0 \text{ and } f_{ik} - \Phi_i \text{ arbitrary}$$

or

$$p_{ik} \neq 0 \text{ and } f_{ik} - \Phi_i = 0 .$$

Here, gradient systems are also of great interest.

Let be given the homogeneous differentiable potential function

$$V(\{tx_{ik}\}) = t^s V(\{x_{ik}\}) \text{ of degree } s > 0 \ .$$

If the fitness functions f_{ik} are defined by the Euclidean gradients

$$f_{ik} = \frac{\partial V}{\partial x_{ik}} \text{ with } \frac{\partial f_{ik}}{\partial x_{jl}} = \frac{\partial f_{jl}}{\partial x_{ik}}$$

then we have between the potential function and the fitness the relation

$$\Phi = \sum_i \Phi_i = \sum_i \sum_k x_{ik} f_{ik} = s \sum_i \sum_k x_{ik} \frac{\partial V}{\partial x_{ik}}$$

and therefore

$$\Phi = sV .$$

4.1. Fixed Points and Saddle Points

Fixed points of the diversified replicator equation are characterized in general by

$$p_{ik} = 0 \text{ and } f_{ik} - \Phi_i \text{ arbitrary}$$

or

$$p_{ik} \neq 0 \text{ and } f_{ik} - \Phi_i = 0.$$

We want to show now that for gradient systems the fixed points p_{ik} are identical to the saddle points of the mean fitness, and the potential °unction taking the simplex condition

$$x_{ik} \in S_m^i$$

into account.

The corresponding Lagrangian is given by

$$L = V - \sum_i l_i \left(\sum_k x_{ik} - 1 \right) - \sum_{\substack{i \; k \\ x_{ik} \in bnd(S_m^i)}} \sum t_{ik} x_{ik}.$$

Then the relation

$$\frac{\partial L}{\partial x_{ik}} = \frac{\partial V}{\partial x_{ik}} - l_i - \underset{x_{ik} \in bnd(S_m^i)}{t_{ik}} = 0$$

holds, and furthermore

$$\sum_k x_{ik} \frac{\partial L}{\partial x_{ik}} = \sum_k x_{ik} \frac{\partial V}{\partial x_{ik}} - l_i - \sum_{\substack{k \\ x_{ik} \in bnd(S_m^i)}} t_{ik} x_{ik} = 0.$$

Because of

$$\sum_{\substack{k \\ x_{ik} \in bnd(S_m^i)}} t_{ik} x_{ik} = 0$$

we have

$$l_i = \sum_k x_{ik} \frac{\partial V}{\partial x_{ik}}.$$

Hence, the relations characterizing saddle points of the fitness function, and the potential function are given by

$$x_{ik}=0 \text{ and } t_{ik}=\frac{\partial V}{\partial x_{ik}}-\sum_k x_{ik}\frac{\partial V}{\partial x_{ik}} \text{ for } x_{ik}\in \operatorname{bnd}(S_m^i)$$

$$x_{ik}\neq 0 \text{ and } \frac{\partial V}{\partial x_{ik}}-\sum_k x_{ik}\frac{\partial V}{\partial x_{ik}}=0 \text{ for } x_{ik}\notin \operatorname{bnd}(S_m^i).$$

But these are exactly the relations for fixed points of the diversified replicator equation. Hence, saddle points of the fitness are identical to the fixed points of the diversified replicator equation under the given conditions.

4.2. The Maximum Principle

Within this section we want to show two important features of the diversified replicator equation.

Just as in the case of the replicator equation there exists for the diversified replicator equation a fundamental theorem, i.e. the mean fitness does not decrease in time.

The second property reflects the feature that under certain conditions the orbits run into the direction of a maximal increase of the mean fitness. At a first glance this seems to be a somewhat strange property because a maximal increase of the mean fitness means that the directions of the orbits are orthogonal to the constant level set of the mean fitness. This is in general not the case.

But it can be shown that for a proper definition of orthogonality this will hold. Such a definition of a vector product was given by Shahshahani (1979) . In this sense a maximum principle holds for the diversified replicator equation. In this section we follow the proof of the maximum principle for replicator equations as it was presented by Sigmund (1985). It can be directly generalized to diversified replicator equations.

The state space of the diversified replicator equation is given by the direct sum of the simplices of the different replicator types

$$S=\bigoplus_i S_m^i.$$

Now we are interested in angles between vectors belonging to the tangent space T_pS of the state space at the point p in int(S). Because of the decomposibility of S into disjoint subspaces we can reduce the analysis to such a subspace.

Then the sum of the components of two vectors of the tangent space

$$T_{p^i}S^i_m$$

at the point

$$p^i=(p_{i1},\dots,p_{im}) \ , \ p^i\in \mathrm{int}(S^i_m)$$

is zero.

Following Shahshahani (1978) we define for two vectors

$$x^i=(x_{i1},\dots,x_{im}) \ , \ y^i=(y_{i1},\dots,y_{im})$$

of the tangent space

$$T_{p^i}S^i_m$$

the vector product

$$(x^i,y^i)_{p^i}=\sum_k \frac{x_{ik}y_{ik}}{p_{ik}}.$$

This vector product is different to the Euclidean vector product

$$(x^i,y^i)=\sum_k x_{ik}y_{ik}$$

because of the factors $1/p_{ik}$. Therefore the kth-term in the sum increases in importance the smaller p_{ik} is.

This definition leads to an orthogonality which depends on p^i, and to a distance which differs from the Euclidean one in that changes near the boundary of the simplices will have larger weights.

If we consider two vectors

$$x=(x^1,\dots,x^n)\quad,\quad y=(y^1,\dots,y^n)$$

of the tangent space T_pS with

$$p=(p^1,\dots,p^n)\quad,\quad p\in\text{int}(S)$$

we have

$$(x,y)=\sum_i\sum_k\frac{x_{ik}y_{ik}}{p_{ik}}.$$

We consider now a differentiable function V mapping a region U in S to R. Then the differential $DV(p)$ is a linear mapping of the tangent space to R for every p in int(S), and there exists a unique vector $gradV(p)$ with

$$(grad\,V(p),y)=DV(p)(y)\quad\forall\quad y\in T_pR^{m\cdot n}\quad.$$

This Euclidean gradient $gradV(p)$ has the components

$$\left(\frac{\partial V}{\partial x_{ik}}\right)(p).$$

On the other hand there exists a unique vector $GradV(p)$ such that

$$(Grad\,V(p),y)_p=DV(p)(y)\quad\forall\quad y\in T_pS\quad.$$

This vector will be denoted as the Shahshahani gradient of V.

With the fitness functions f_{ik} we connect now a vector field on S with the coordinates

$$\hat{f}_{ik}=x_{ik}\left[f_{ik}(x)-\Phi_i(x)\right]$$

with

$$\Phi_i=\sum_k x_{ik}f_{ik}\ \text{and}\ \Phi=\sum_i\Phi_i.$$

Thus, we have the diversified replicator equation in the form

$$\dot{x}=\hat{f}(x).$$

The following two properties can be checked easily:

(i) The state space

$$S = \oplus_{l} S_m^l$$

and the faces of the simplices

$$S_m^l$$

are invariant with respect to the diversified replicator equation.

(ii) If

$$f \text{ and } g$$

are two equivalent vector fields then

$$\hat{f} = \hat{g} \text{ on } S$$

holds. Here

$$f \sim g \text{ holds if } f_{lk}(x) - g_{lk}(x)$$

are independent of k for all x in S.

Now we can characterize the diversified replicator equations which are Shahshahani gradients:

$$\hat{f} = Grad V \text{ iff } f \sim grad V \ .$$

We assume

$$f \sim grad V \ .$$

For

$$g \sim h \text{ holds } \hat{g} = \hat{h}$$

on S. Without loss of generality we can assume that

$$f = grad V \ .$$

This yields for y in T_pS

$$(\mathfrak{f}(p),y)_p = \sum_i \sum_k \frac{1}{p_{ik}} p_{ik}(f_{ik} - \Phi_i) y_{ik}$$

$$(\mathfrak{f}(p),y)_p = \sum_i \sum_k f_{ik} y_{ik} - \sum_i \Phi_i \sum_k y_{ik}$$

$$(\mathfrak{f}(p),y)_p = \sum_i \sum_k f_{ik} y_{ik}$$

because of

$$\sum_k y_{ik} = 0$$

and therefore

$$(\mathfrak{f}(p),y)_p = \sum_i \sum_k \frac{\partial V}{\partial x_{ik}} y_{ik} = DV(p)(y).$$

Hence

$$\mathfrak{f} = Grad\, V$$

holds.

If

$$\mathfrak{f} = Grad\, V$$

then we have

$$\sum_i \sum_k f_{ik} y_{ik} = \sum_i \sum_k \frac{\partial V}{\partial x_{ik}} y_{ik}$$

for all y in T_pS. This implies for

$$y_{ik} = 1 \quad , \quad y_{im} = -1 \text{ and } y_{il} = 0 \quad \forall\, l \neq k,m$$

that

$$\sum_i \frac{\partial V}{\partial x_{ik}}(p) - \sum_i \frac{\partial V}{\partial x_{im}}(p) = \sum_i f_{ik}(p) - \sum_i f_{im}(p).$$

It follows that

$$\sum_i \frac{\partial V}{\partial x_{ik}}(p) - \sum_i f_{ik}(p)$$

is independent of k, and therefore

$$f \sim \operatorname{grad} V \quad .$$

If f is a Euclidean gradient the orbits of

$$\dot{x} = f(x)$$

are orthogonal in the sense of Shahshahani to the constant level set of V on S.

Let the potential function V be a homogeneous function of degree s, i.e.

$$V(\{t x_{ik}\}) = t^s V(\{x_{ik}\})$$

for all t in R. Then the mean fitness fulfils the relation

$$\Phi(p) = \sum_i \sum_k p_{ik} f_{ik}(p) = \sum_i \sum_k p_{ik} \frac{\partial V}{\partial x_{ik}}(p)$$

and finally

$$\Phi(p) = sV(p).$$

Hence the mean fitness increases in the sense of Shahshahani at a maximal rate. The increase is given by

$$\dot{\Phi}(x) = s\dot{V}(x) = s \sum_i \sum_k \frac{\partial V}{\partial x_{ik}} \dot{x}_{ik}$$

$$\dot{\Phi}(x) = s \sum_i \sum_k \frac{\partial V}{\partial x_{ik}} x_{ik} \left[\frac{\partial V}{\partial x_{ik}} - \sum_l x_{il} \frac{\partial V}{\partial x_{il}} \right]$$

$$\dot{\Phi}(x) = s \sum_i \left\{ \sum_k x_{ik} [f_{ik}(x)]^2 - \left[\sum_l x_{il} f_{il}(x) \right]^2 \right\} \geq 0.$$

Up to the factor s, the increase of the fitness is the sum of variances of random variables assuming the values $f_{ik}(x)$ with probability x_{ik}, $k = 1,\dots,m$, for all replicator types $i = 1,\dots,n$. This is the fundamental theorem for diversified replicator equations.

Because of the fact that the state space S is the direct sum of the single simplices, these features hold for every such simplex.

Without loss of generality we fix all values x_{ik}, i.e. $dx_{ik}/dt = 0$, for $i = 2,\dots,n$ and all k.

Then the increase of the mean fitness is dependent only on x_{1k}, and we have

$$\dot{\Phi}(x) = s\dot{V}(x) = s\sum_k \frac{\partial V}{\partial x_{1k}}\dot{x}_{1k}$$

$$\dot{\Phi}(x) = s\left\{\sum_k x_{1k}[f_{1k}(x)]^2 - \left[\sum_l x_{1l} f_{1l}(x)\right]^2\right\} \geq 0.$$

This feature is advantageous for the computation of fixed points.

4.3. Fitness on Graphs

Within this section we want to establish a connection between edge valued graphs and the mean fitness of the diversified replicator equation.

For this reason we consider a graph

$$G = (V, E)$$

with the vertex set

$$V = \{V_i\} \quad , \quad i = 1,\dots,n$$

and the edge set

$$E = \{E_{ij}\} \quad , \quad i,j = 1,\dots,n \quad .$$

To every edge we assign a value G_{ij} and to every vertex a vector

$$x^i = (x_{i1},\dots,x_{im}) \in S^i_m.$$

Furthermore we consider a graph

$$G^* = (V^*, E^*)$$

with the vertex set

$$V^* = (V_k^*) \quad , \quad k = 1, \ldots, m$$

and the edge set

$$E^* = (E_{kl}^*) \quad , \quad k, l = 1, \ldots, m \quad .$$

To every edge we assign a value

$$C_{kl} = \sum_i x_{ik} \sum_j G_{ij} x_{jl}$$

and to every vertex a value

$$X_k = \sum_i x_{ik} .$$

These vertices constitute a kind of super vertices. The values of the super vertices we get from the corresponding proportions of the vertex values of the graph G. x_{ik} represents the probability with which the vertex i of the graph G belongs to a super vertex k of the graph G^*. x_{ik} represents the probability with which vertex i of the graph G belongs to a super vertex k of the graph G^*.

Imposing now extremality demands upon the edge values

$$C_{kl} = \sum x_{ik} \sum_j G_{ij} x_{jl}$$

generates characteristic structures of the graph

$$G^* = (V^*, E^*) \quad .$$

So it is possible to generate a fuzzy hypercycle corresponding to Figure 4.3.1 if we demand

$$C_{kl} \rightarrow \max \text{ for } l = k-1 \pmod{m} \ .$$

Fuzzy means here that the directed edge weights of adjacent super vertices should be as large as possible, and all other edge weights should be as small as possible.

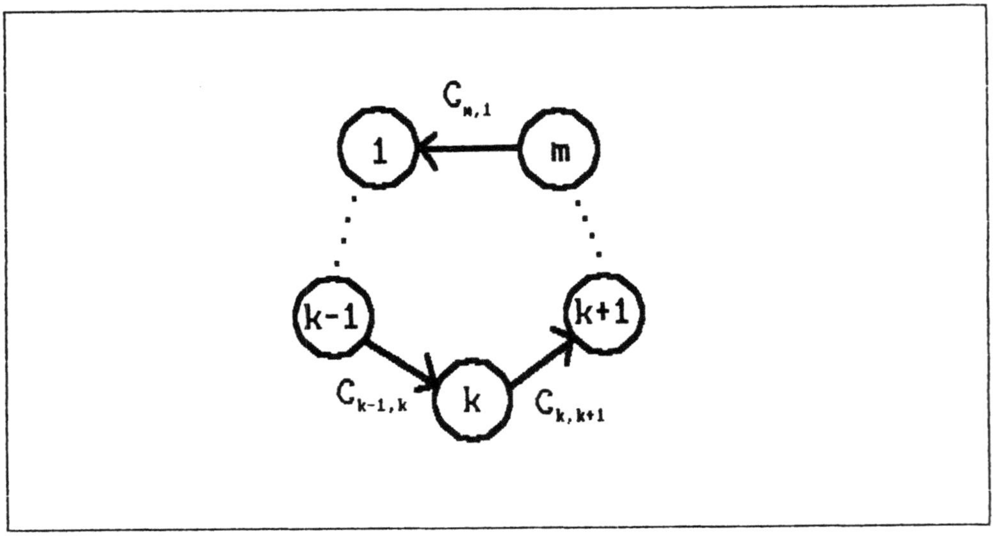

Figure 4.3.1 Fuzzy hypercycle

Usually, demands as mentioned above will lead to vector optimization problems. Hence we have to consider not only one optimal solution but a set of efficient solutions. Within this set we finally have to choose a solution with the highest preference. If preferences are a priori at hand it is possible to aggregate the demands into a so-called utility function.

For the above mentioned example we may raise the demand for a utility function of the form

$$V = \prod_k C_{kl} \rightarrow \max \text{ for } l = k-1 \pmod{m} \ .$$

Such a demand means that the edge values C_{kl} should be maximal, and should have at the same time approximately equal values. We will meet such demands e.g. in connection with partitioning problems later on.

Minimizing demands can be formalized in much the same way.

In general we can consider a utility function as a potential function V with

$$V = V(\{C_{kl}\})$$

over a selected set of edge weights

$$C_{kl} = \sum_i x_{ik} \sum_j G_{ij} x_{jl} \quad , \quad k \in K \quad , \quad l \in L$$

with K, L subscript sets.

Let V be a differentiable homogeneous function of degree $s > 0$. Then we have the Euclidean gradients

$$\frac{\partial V}{\partial x_{ik}} = \sum_l \frac{\partial V}{\partial C_{kl}} \cdot \frac{\partial C_{kl}}{\partial x_{ik}}$$

and with

$$\frac{\partial C_{kl}}{\partial x_{ik}} = \sum_j G_{ij} x_{jl} = (Gx_{.l})_i$$

the relation

$$\frac{\partial V}{\partial x_{ik}} = \sum_l \frac{\partial V}{\partial C_{kl}} (Gx_{.l})_i \quad .$$

Then the fitness of the replicator type i is given by

$$\Phi_i = s \sum_k x_{ik} \sum_l \frac{\partial V}{\partial C_{kl}} (Gx_{.l})_i$$

and the total fitness by

$$\Phi = s \sum_i \sum_k x_{ik} \sum_l \frac{\partial V}{\partial C_{kl}} (Gx_{.l})_i$$

$$\Phi = s \sum_k \sum_l \frac{\partial V}{\partial C_{kl}} C_{kl} = sV \quad .$$

Since the mean fitness is monotone increasing as mentioned in the previous section, it is possible by means of the diversified replicator equation to compute optimal structures

$$V \to \max \text{ of the graph } G^{*} \text{ over } G \quad .$$

In case of the fuzzy hypercycle we have for the diversified replicator equation the fitness functions

$$f_{lk} = \frac{\partial V}{\partial x_{lk}} = V \frac{(Gx_{.l})_l}{C_{kl}}$$

as well as

$$\Phi_l = \sum_k x_{lk} f_{lk} \text{ and } \Phi = \sum_l \sum_k x_{lk} f_{lk}$$

for $l = k-1(\mathrm{mod}\ m)$.

4.4. Homogeneous Replication Rate Control

Usually the potential function

$$V(\{C_{kl}\})$$

has a large number of local maxima. But we want to find a global maximum, or at least a solution near the global maximum.

Furthermore it should be noted corresponding to the previous considerations that there exist no unique assignments of the vertices of the graph G to the vertices of the graph G^*. The assignments are fuzzy or probabilistic ones. We consider the advantages of fuzzy assignments for the solution of practically important problems in Section 6. In case of the usual notion of combinatorial optimization problems a unique assignment is demanded.

To take care to these demands we introduce a homogeneous replication rate control as selection pressure as in the case of the replicator equation. The edge values of the graph G will be modified by the control u

$$G_u := G_u + u \quad .$$

A graph modified in such a way has the structure depicted in Figure 4.4.1.

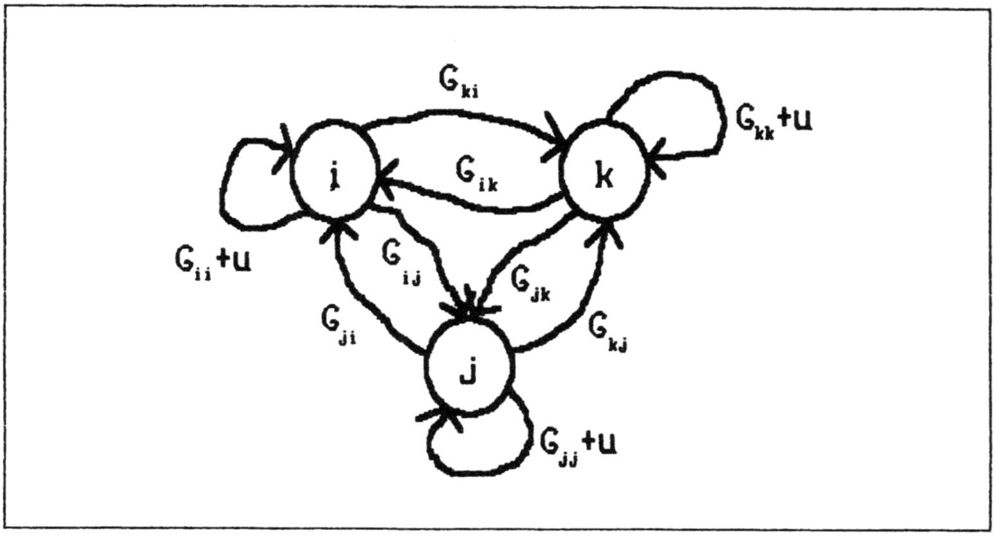

Figure 4.4.1 By the selection pressure *u* modified graph *G*

Corresponding to Section 4.3. with

$$X_{kl} = \sum_{i} x_{il} x_{ik}$$

the potential function is given by

$$V = V(\{C_{kl} + uX_{kl}\}) \quad .$$

To consider positive and negative values of the selection pressure we use

$$u = sign(u)|u|$$

with

$$sign(u) = +1 \text{ for } u \geq 0$$

$$sign(u) = -1 \text{ for } u < 0$$

Then the potential function can be written as

$$V = V\left(\left\{|u|\left[\frac{C_{kl}}{|u|} + sign(u)X_{kl}\right]\right\}\right) \quad .$$

If we assume that the potential function is homogeneous of degree $r > 0$

with respect to the selection pressure u we have

$$V = |u|^r V\left(\left\{\frac{C_{kl}}{|u|} + sign(u) X_{kl}\right\}\right) .$$

As fitness function f_{ik} we use the Euclidean gradients

$$f^*_{ik} = \frac{\partial V}{\partial x_{ik}} = |u|^r \sum_l g_{kl}\left[\frac{(Gx_{.l})_i}{|u|} + sign(u) x_{il}\right]$$

$$f^*_{ik} = |u|^r f_{ik}$$

where the g_{kl} are the partial differentials

$$g_{kl} = \frac{\partial V\left(\left\{\frac{C_{kl}}{|u|} + sign(u) X_{kl}\right\}\right)}{\partial\left(\frac{C_{kl}}{|u|} + sign(u) X_{kl}\right)} .$$

Then the fitness functions of the replicator types are given by

$$\Phi^*_i = \sum_k x_{ik} f^*_{ik} = |u|^r \sum_k x_{ik} f_{ik}$$

$$\Phi^*_i = |u|^r \Phi_i$$

and the diversified replicator equation by

$$\dot{x}_{ik} = x_{ik} |u|^r (f_{ik} - \Phi_i) .$$

Considering $|u|^r$ as a time transformation we may omit it, and finally we have

$$\dot{x}_{ik} = x_{ik}(f_{ik} - \Phi_i) .$$

We will analyze now the two extreme cases for very large and very small selection pressure. In these two cases the structure of the graph will be completely destroyed, i.e. the edge weights G_{ij} do not play any role.

How is this reflected by the potential function V over the graph G^*?

For very large and very small values of the selection pressure u we have for the fitness functions the following terms

$$\lim_{u\to\pm\infty} g_{kl} = \hat{g}_{kl} = \frac{\partial V((sign(u)X_{kl}))}{\partial(sign(u)X_{kl})}$$

$$\lim_{u\to\pm\infty} f_{ik} = \hat{f}_{ik} = sign(u)\sum_{l} \hat{g}_{kl} x_{il}$$

$$\lim_{u\to\pm\infty} \Phi_i = sign(u)\sum_{k} x_{ik} \sum_{l} \hat{g}_{kl} x_{il}$$

$$\lim_{u\to\pm\infty} \Phi = \hat{\Phi} = sign(u)\sum_{i}\sum_{k} x_{ik} \sum_{l} \hat{g}_{kl} x_{il} \quad .$$

Because of the fundamental theorem this yields

$$\sum_{i}\sum_{k} x_{ik} \sum_{l} \hat{g}_{kl} x_{il} \to \max \text{ for } u \to +\infty$$

$$\sum_{i}\sum_{k} x_{ik} \sum_{l} \hat{g}_{kl} x_{il} \to \min \text{ for } u \to -\infty.$$

As an analogue to the replicator equation we have a fixed point in the interior of the single simplices because of the relation

$$\hat{f}_{ik} - \hat{\Phi}_i = 0$$

with the matrix

$$\hat{G} = \begin{pmatrix} \hat{g}_{11} & \cdots & \hat{g}_{1m} \\ \cdots & \cdots & \cdots \\ \hat{g}_{m1} & \cdots & \hat{g}_{mm} \end{pmatrix} \text{ and } x_{i.} = \begin{pmatrix} x_{i1} \\ \cdots \\ x_{im} \end{pmatrix} , \quad e = \begin{pmatrix} 1 \\ \cdots \\ 1 \end{pmatrix}$$

given by

$$x_{i.} = \frac{adj(\hat{G})e}{(e, adj(\hat{G})e)} \quad .$$

Because of these equations a uniform distribution of the x_{ik} to all super vertices of the graph G^* for very small selection pressure is characterized by the conditions concerning the partial differentials of the potential function

$$[adj(\hat{G})e]_k = [adj(\hat{G})e]_l$$

for all k and l.

If these conditions are fulfilled a stable fixed point occurs at the barycentres of the simplices for for very small values of u.

Conversely we have stable fixed points for very large values of the selection pressure u only at vertices of the single simplices.

As in the case of the replicator equation the scaling of the global state depends on the selection pressure. For very small selection pressure u this yields the global state of uniform cooperation, and for very large selection pressure u the global state of uniform competition.

This type of selection pressure control is especially used for graph partitioning algorithms because in this case only the edge values C_{kk} have a dominating influence on the evolution process. Here, no additive influence of the probabilities x_{il}, $l = 1, \ldots$, on the self-reproduction rate of replicator type i in area k occurs.

To avoid such cross influences for other applications as e.g. the module placement problem and the travelling salesman problem we introduce a selection pressure control in such a way that its influence on x_{ik} depends only on this x_{ik}.

For the potential function

$$V = V(\{C_{kl}\})$$

we have the Euclidean gradients

$$f_{ik} = \frac{\partial V}{\partial x_{ik}} = \sum_l \frac{\partial V}{\partial C_{kl}} \cdot (Gx_{.l})_i \quad .$$

Using the f_{ik} as fitness functions and introducing a selection pressure control u according to

$$f_{ik}^{*} := f_{ik} + u x_{ik}$$

yields the diversified replicator equation

$$\dot{x}_{ik} = x_{ik}\left(f_{ik} + u x_{ik} - \Phi_i\right)$$

with the mean value of the fitness of alleles i

$$\Phi_i = \sum_k x_{ik} f_{ik} + u \sum_k x_{ik}^2 \quad .$$

For this equation we have with respect to the selection pressure control the following important invariance property:

The diversified replicator equation

$$\dot{x}_{ik} = x_{ik}\left(f_{ik} - u\left(1 - x_{ik}\right) - \Phi_i\right)$$

with the mean fitness of alleles i

$$\Phi_i = \sum_k x_{ik} f_{ik} - u\left(1 - \sum_k x_{ik}^2\right)$$

has the same flow on the simplices as the former equation because the constant selection pressure term can be cancelled.

Let us assume that all fitness functions f_{ik} are nonnegative. Then all terms of the diversified replicator equation with the selection pressure control $u x_{ik}$ are nonnegative for u nonnegative. On the other hand, for the selection pressure control $u(1-x_{ik})$ all terms of the diversified replicator equation are nonnegative for negative u.

Though this property does not hold in general for the recurrence diversified replicator equation, it holds for fixed points. We therefore apply these controls to ensure simplex invariance in the following way:

$$x'_{ik} = x_{ik} \frac{f_{ik} + u x_{ik}}{\Phi_i} \quad \text{for } u \geq 0$$

$$x'_{ik} = x_{ik} \frac{f_{ik} - u\left(1 - x_{ik}\right)}{\Phi_i} \quad \text{for } u < 0 \quad .$$

For the two extreme cases of very small and very large selection pressure u these equations show the same behavior as the replicator equation (see Section 3.3.).

Connected to these considerations we obtain a very interesting fact.

The control terms can be thought of as first order approximations of the functions

$$u\ln(x_{ik}) = -u(1-x_{ik}) + \ldots \text{ for } u<0$$

$$u(1+\ln(x_{ik})) = ux_{ik} + \ldots \text{ for } u\geq 0 \quad .$$

For this case the fitness functions of alleles i are given by

$$\Phi_i = \sum_k x_{ik}f_{ik} + u\sum_k x_{ik}\ln(x_{ik}) \text{ for } u<0$$

$$\Phi_i = \sum x_{ik}f_{ik} + u(1+\sum_k x_{ik}\ln(x_{ik}) \text{ for } u\geq 0 \quad .$$

The second term on the right-hand side of these equations are entropy measures. This means, the evolution process proceeds corresponding to a replicator dynamics and is controlled via a selection pressure based on the entropy of the population.

This type of control has to be analyzed in the future in more detail. It represents a very promising way of combining models of biological evolution and of statistical mechanics.

5. Replicator Network Algorithms

Within this section we want to demonstrate by means of three concrete problems the application and efficiency of the diversified replicator equation for the solution of combinatorial optimization problems.

For these concrete applications we use the diversified recurrence replicator equation

$$x'_{ik} = x_{ik} \frac{f_{ik}}{\Phi_i}$$

because of the advantages implementing it on a digital computer. As we already saw this equation has the same fixed points we are interested in as the continuous diversified replicator equation.

The control of the evolution process will be realized by the selection pressure control u. The control schedule will be chosen in the following way: We start at a low selection pressure, i.e. u sufficiently small, increasing it very slowly.

For every fixed value of the selection pressure the diversified replicator equation will be iterated until a relative error threshold of the fitness function

$$\frac{|\Phi' - \Phi|}{\Phi} \leq e$$

will be reached.

Of course other meaningful indicators may be used as e.g. the relative change of the information entropy

$$S(x) = -\sum_i \sum_k x_{ik} \log(x_{ik})$$

or of the mean value

$$E(x) = \sum_i \sum_k x_{ik}^2$$

which are both monotone functions in u.

To avoid a resting of the algorithms at an unstable fixed point or at faces of the simplices for every new generation (iteration) mutants of the replicator types within the different " areas" will be generated randomly. This will be done by a mutation rate control h, $h > 0$.

Let Y be a uniformly distributed random variable with the realizations y_{ik}. The initial distribution of the x_{ik} for fixed u will be randomly disturbed according to

$$x_{ik} := \frac{x_{ik} + h y_{ik}}{1 + h \sum_k y_{ik}}.$$

Other methods for generating mutants are of course possible. The above mentioned method worked very well for the following applications.

The evolution process proceeds as follows:

1 **Start** with a uniform distribution of the x_{ik}
2 Set the selection pressure u sufficiently low
3 Disturb the x_{ik} slightly (**mutation**)
4 Iterate the recurrence equation (**replication and selection**)
5 If the probabilities x_{ik} are stabilized go to 6, else go to 3
6 If the assignments are sufficiently crisp then **Stop**
7 Increase the selection pressure u slightly and go to 3

The description of the applied control mechanisms is now complete.

A basic task in solving combinatorial optimization problems consists in the formulation of the fitness functions. The suitable formulation of the optimization goals is essential for all optimization problems and methods.

We want to explain the method including the formulation of suitable fitness functions by means of three optimization problems, the partitioning of undirected graphs, the placement of modules on a grid, and the travelling salesman problem.

5.1. The Graph Partitioning Problem

Because of different reasons the partitioning of graphs is important for the layout design of integrated circuits and printed circuit boards. Main reasons are:

(i) A partitioning of the connection structure can be used to make subsequent design steps possible by lowering the complexity of the problems for the single partitions.

(ii) If there is given a predefined basic structure then the given graph has to be embedded into this structure.

Case (ii) refers e.g. to gate arrays, a technology which makes an economic realization of application specific integrated circuits possible.

For gate arrays the so-called master wafers with a given structure will be produced in large numbers. The structure realizing the application specific function will then be imposed by only a few process steps to the master wafer.

A typical example is the CMOS gate array system U 5200 of Carl Zeiss Jena.

5.1.1. Fitness

For graph partitioning usually a variety of conditions and constraints has to be considered. They depend on the concrete applications. They also require the formulation of appropriate fitness functions.

One demand is in common to all partitioning problems: The vertices of the graph G should be assigned to the super vertices of the graph G^* in such a way that the edge weights C_{kl}, k unequal to l, between super vertices should be as small as possible, and the edge weights C_{kk} within a super vertex should be as large as possible under the given constraints.

This demand yields the aspired basic structure of G^* shown in Figure 5.1.1. It is characterized by unconnected super vertices.

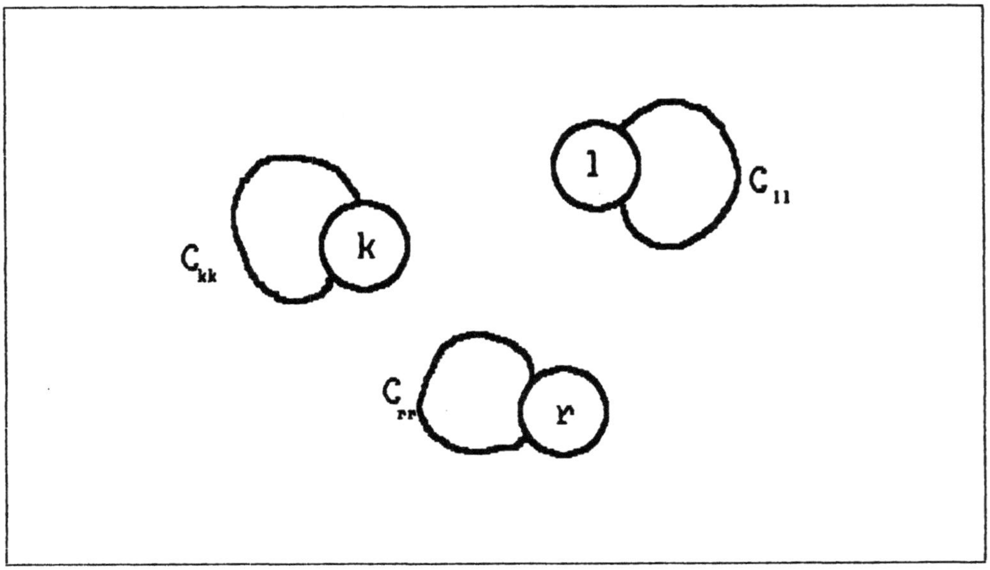

Figure 5.1.1 Basic structure of the graph G^* for the graph partitioning problem

The associated vector optimization problem is given then by

$$C_{kk} \to \max \quad , \quad x_{ik} \in \{0, 1\}.$$

The inclusion of further demands and constraints as

(i) relationship of the the sizes of the C_{kk}

(ii) relationship of the sizes of the X_k

(iii) fixed or variable number of super vertices

(iv) consideration of neighbourhood relations between vertices of the graph G because of appended attributes (e.g. electrical properties in case of IC design)

is decisive for the further steps. Because of these further constraints, though the above listing is not complete, one can imagine the large variety of possible fitness functions.

Here we want to present a fitness function, or potential function,

$$V=\prod_k (C_{kk}+a) \quad , \quad k=1,\ldots,m<n$$

which is rendered because of its flexibility. It includes especially constraints (i) and (iii). The flexibility is characterized by

(i) The number of super vertices can be fixed.

(ii) For $a = 0$, the fitness function is equal to the product of the edge weights C_{kk} of the super vertices. This implies that the edge weights C_{kk} should have approximately equal size.

(iii) For increasing values of $a > 1$, the sum of the edge weights of the super vertices has an increasing influence because of

$$\prod_k (C_{kk}+a)=\prod_k C_{kk}+\ldots+a^{m-1}\sum_k C_{kk}+a^m .$$

(iv) For increasing values of $a > 1$, the number of super vertices adapts more and more to the structure of the graph G.

In Section 5.1.2. we want to show these features by means of an example. In Section 5.1.3. we compare this approach with other methods using a somewhat larger industrial example.

Starting with the above formulated fitness function of the total gene pool we have the fitness functions of the different replicator types i within the "areas" k by

$$f_{ik}=\frac{\partial V}{\partial x_{ik}}=2V\frac{(Gx_{.k})_i}{C_{kk}+a}$$

and the total fitness of replicator type i by

$$\Phi_i=\sum_k x_{ik}f_{ik}=2V\sum_k x_{ik}\frac{(Gx_{.k})_i}{C_{kk}+a} .$$

The total fitness of the gene pool is then given by

$$\Phi=2V\sum_k\frac{C_{kk}}{C_{kk}+a} .$$

For $a = 0$ we have

$$\Phi = 2mV \quad .$$

For this fitness functions the diversified replicator equation can be written as

$$\dot{x}_{ik} = x_{ik} 2V \left[\frac{(Gx_{.k})_i}{C_{kk}+a} - \sum_l x_{il} \frac{(Gx_{.l})_i}{C_{ll}+a} \right] \quad .$$

For computer implementations we use the recurrence form

$$x'_{ik} = x_{ik} \frac{\dfrac{(Gx_{.k})_i}{C_{kk}+a}}{\sum_l x_{il} \dfrac{(Gx_{.l})_i}{C_{ll}+a}} .$$

It should be noted that the factor $2V$ acting as a time transformation in case of the continuous diversified replicator equation can be cancelled for the recurrence equation. In this case it has no influence.

The selection pressure control acts via the diagonal elements of the interaction matrix. The application will be demonstrated in the next sections.

A fitness function emphasizing especially the constraints (ii) and (iii) is given by

$$V = \sum_k C_{kk} \rightarrow \max \quad , \quad k = 1,\dots,m$$

with the constraints

$$X_k = \sum_i x_{ik} = \text{const} \; .$$

In this case the Euclidean gradients are given by

$$\frac{\partial V}{\partial x_{ik}} = (Gx_{.k})_i \quad .$$

Using the diversified replicator equation we introduce a selection pressure control u such that

$$f_{ik} = \frac{\sum_{j} G_{ij} x_{jk} + u x_{ik}}{X_k} \quad \text{for } u \geq 0$$

$$f_{ik} = \frac{\sum_{j} G_{ij} x_{jk} - u(1 - x_{ik})}{X_k} \quad \text{for } u < 0 \quad .$$

5.1.2. Example

We want to clarify the features of the approach for the given fitness functions by means of the example **TEST#31**.

For two partitions Figure 5.1.2 shows the evolution of the fitness function. The corresponding graph for the edge weights C_{kk} is presented in Figure 5.1.3. The same variables are depicted in Figure 5.1.4 and 5.1.5 for three partitions.

The resulting assignments of the vertices of the graph G to the super vertices (partitions) of G^* dependent on the selection pressure u are shown for two partitions in Figure 5.1.6 and for three partitions in Figure 5.1.7.

In both cases the selection pressure was increased by 0.5 after stabilization with $e < 0.01$. An initial value of $u_0 = -1.5$ was used. From the figures it can be seen quite clear that after an initial cooperative phase, where all values of the C_{kk} increase, an increasing competition between the C_{kk} occurs.

This observation is also reflected by the assignment values x_{ik}. For a sufficiently large selection pressure we have a unique assignment of vertices of the graph G to super vertices of G^*. Furthermore we can see that for $u < 0$ the fitness increases monotonously with u.

Figures 5.1.8 to 5.1.10 show the graphs of the fitness as well as of the edge weights C_{kk} versus the selection pressure u for 2, 3, and 4 partitions. The selection pressure was increased following the schedule $u = -1.0(.1)\ldots$ with an error bound for the iterations of $e = 0.01$. Here it can be seen also that the fitness increases monotonously for $u < 0$. Especially for $m = 4$ partitions we see that the fitness can decrease with increasing selection pressure $u > 0$. This is also an indicator for a increasing competition between replicator types i for the different "areas" k.

The influence of the control variable a which can be considered as a balance measure between an multiplicative and an additive aggregation of the edge weights C_{kk} is depicted in Figures 5.1.12 through 5.1.16 for m = 2 partitions. Especially for $a = 0$ the influence of the error bound e = 0.0001 can be seen. A smaller error bound leads to a stronger emphasis of the fitness function.

It is interesting to see that an almost unique assignment of a vertex to a partition may change rapidly to another partition. This is a typical effect of evolving systems. It becomes very clear for the case of $a = 10$ corresponding to Figure 5.1.17.

Figure 5.1.18 shows the resulting edge weights C_{kk} for different values of a corresponding to Table 5.1.1. From this it becomes clear that all assignment solutions are efficient solutions in the sense of vector optimization

$$C_{kk} \rightarrow \max .$$

But the influence of the control variable a is more fare reaching. If we impose too strong constraints on the solution set then a quite natural adaption of the constraints to the graph structure G occurs. To show this we consider the case of m = 4 partitions. For a = 5 we have the graph presented in Figure 5.1.19. A further increase of a to a = 10 brings about an adaption of the number of the partitions to the real given structure of the graph G. This becomes clear by Figure 5.1.20.

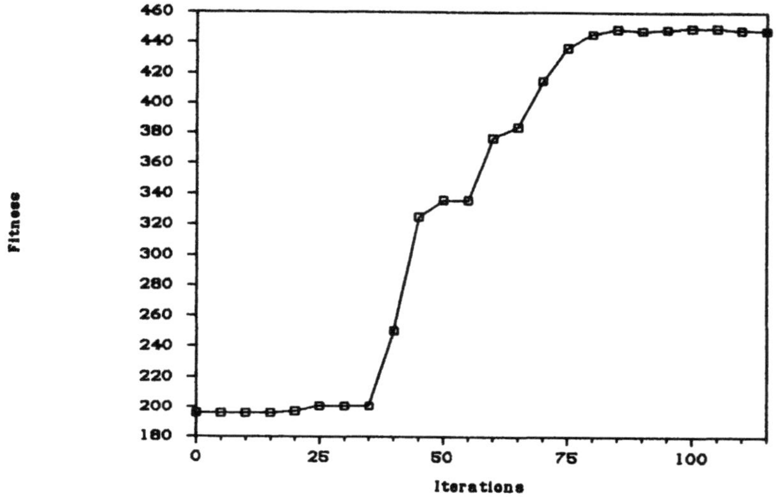

Figure 5.1.2 Graph of the fitness versus iteration count (e = 0.01, h = 0.1, a = 0, u = -1.5(.5)4.5, m = 2)

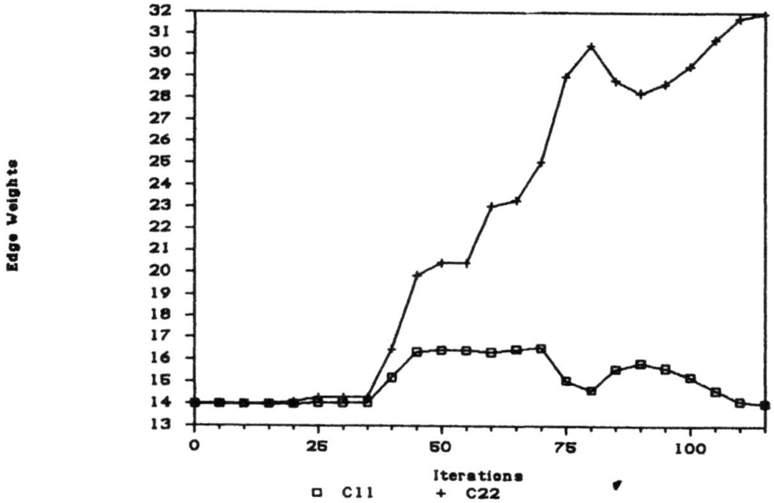

Figure 5.1.3 Edge weights C_{kk} versus iteration count (e = 0.01, h = 0.1, a = 0, u = -1.5(.5)4.5, m = 2)

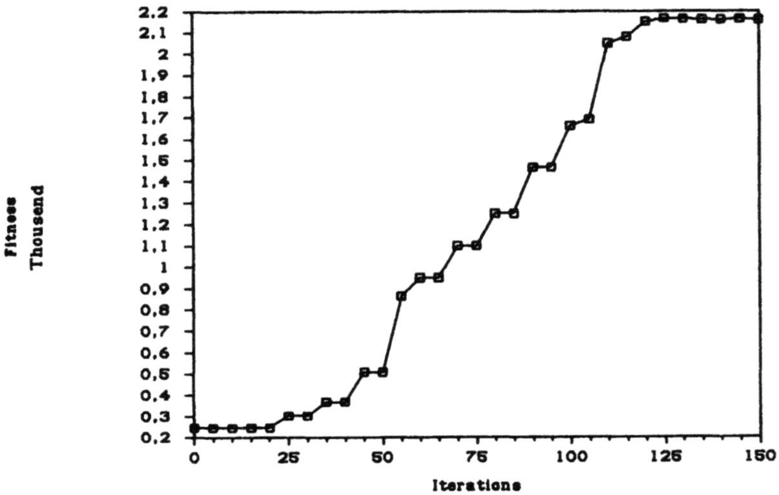

Figure 5.1.4 Graph of the fitness versus iteration count (e = 0.01, h = 0.1, a = 0, u = -1.5(.5)3.5, m = 3)

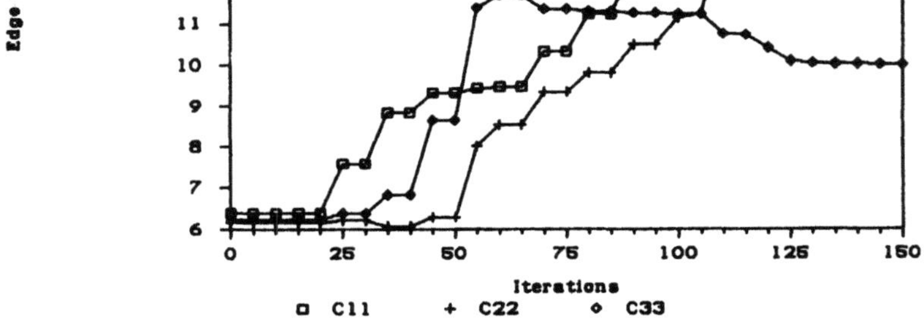

Fig 5.1.5. Graph of C_{kk} versus iteration count (e = 0.01, h = 0.1, a = 0, u = -1.5(.5)3.5, m = 3)

45	54
65	34
34	65
36	63
99	--
99	--
57	42
17	82
16	83
19	80

u = -1.5

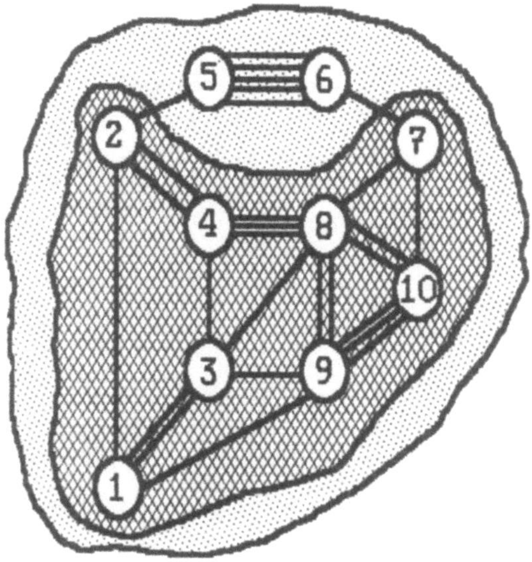

50	49
83	16
32	67
34	65
99	--
99	--
60	39
7	92
3	96
5	94

u = -1.0

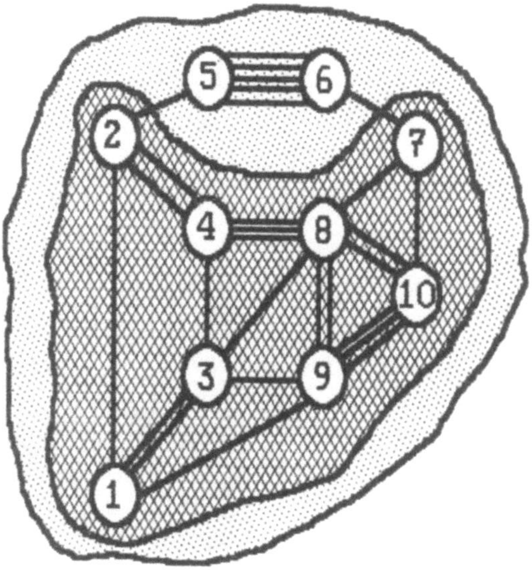

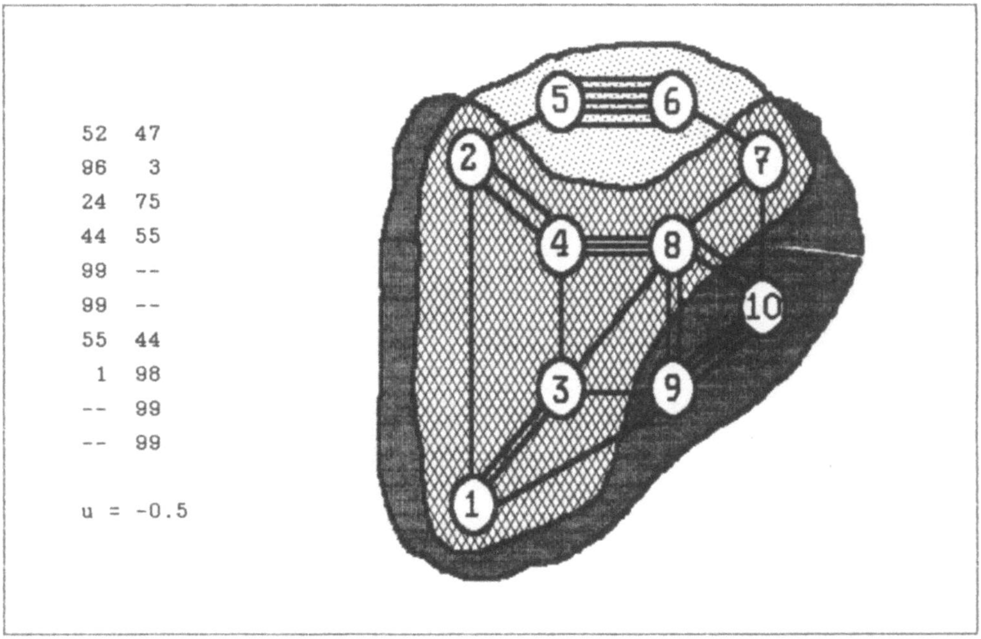
52 47
96 3
24 75
44 55
99 --
99 --
55 44
1 98
-- 99
-- 99
u = -0.5
5
6
2
7
4
8
10
3
9
1

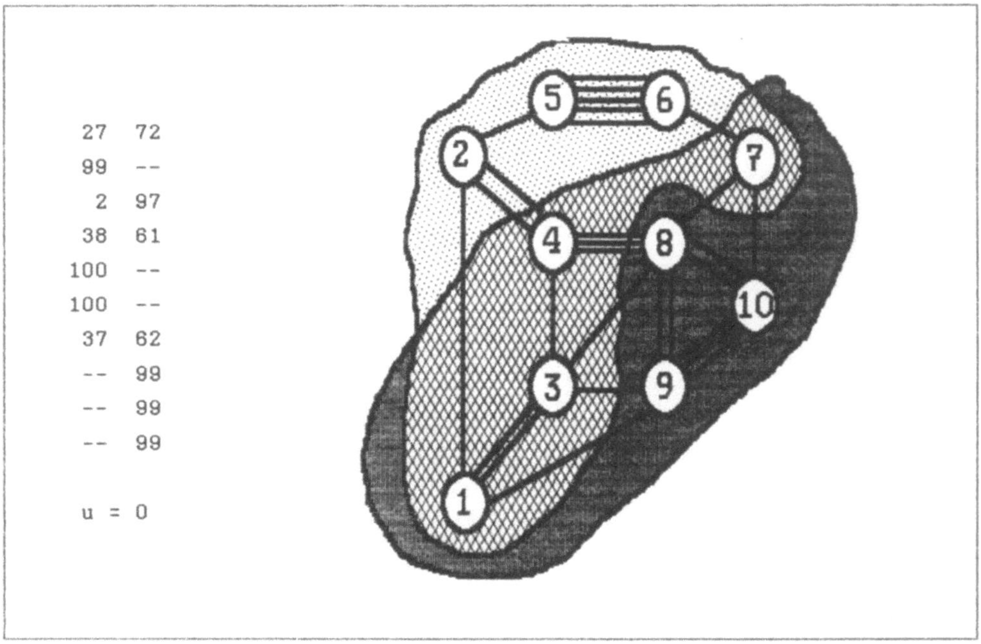
27 72
99 --
2 97
38 61
100 --
100 --
37 62
-- 99
-- 99
-- 99
u = 0
5
6
2
7
4
8
10
3
9
1

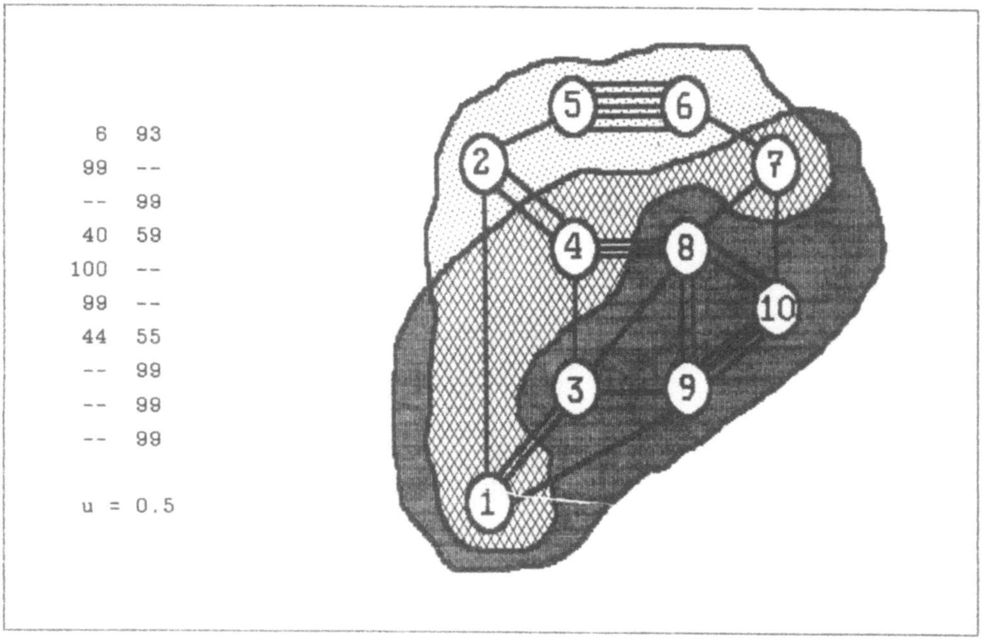
6 93
99 --
-- 99
40 59
100 --
99 --
44 55
-- 99
-- 99
-- 99
u = 0.5
5
6
2
7
4
8
10
3
9
1

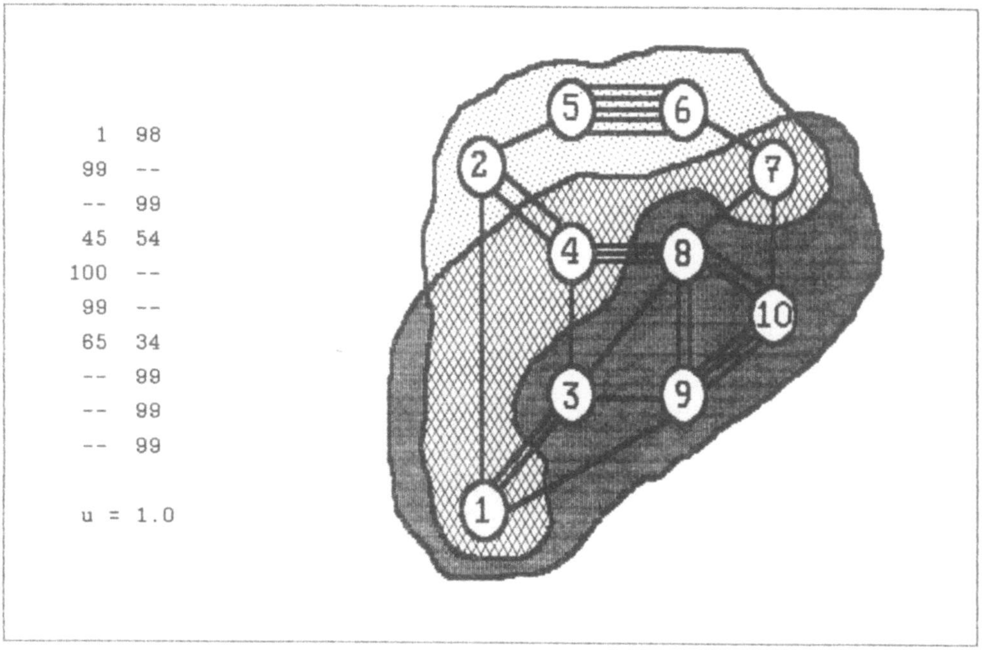
1 98
99 --
-- 99
45 54
100 --
99 --
65 34
-- 99
-- 99
-- 99
u = 1.0
5
6
2
7
4
8
10
3
9
1

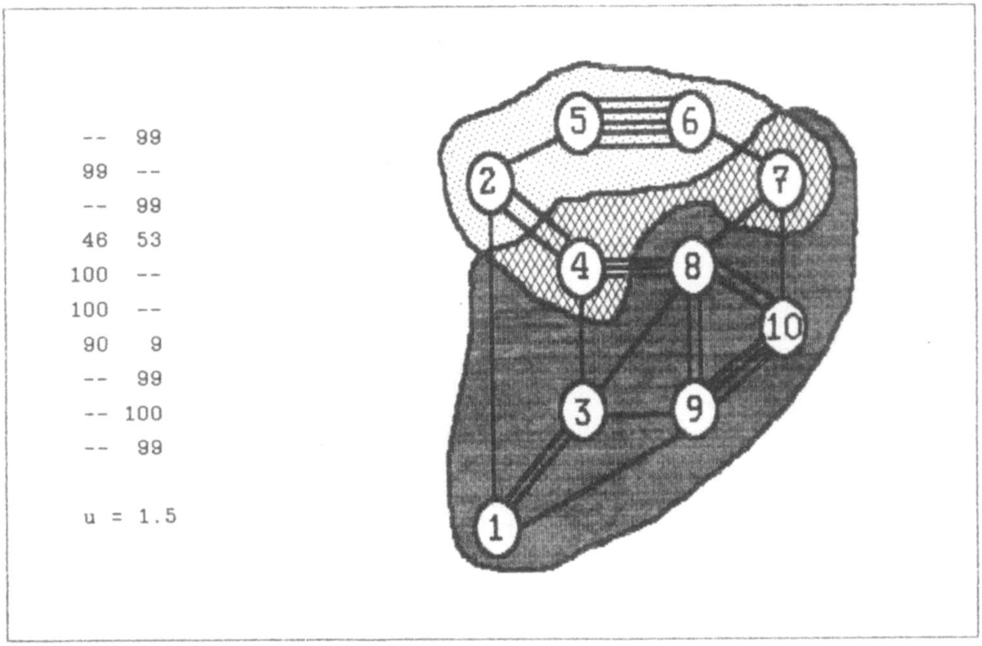

-- 99
99 --
-- 99
46 53
100 --
100 --
90 9
-- 99
-- 100
-- 99
u = 1.5
5
6
2
7
4
8
10
3
9
1

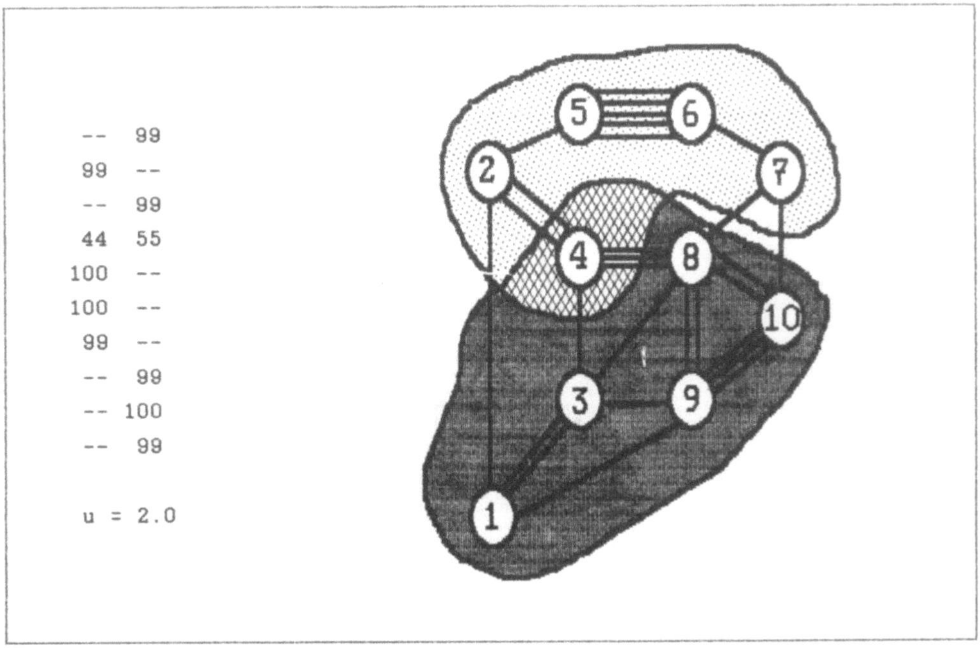

-- 99
99 --
-- 99
44 55
100 --
100 --
99 --
-- 99
-- 100
-- 99
u = 2.0
5
6
2
7
4
8
10
3
9
1

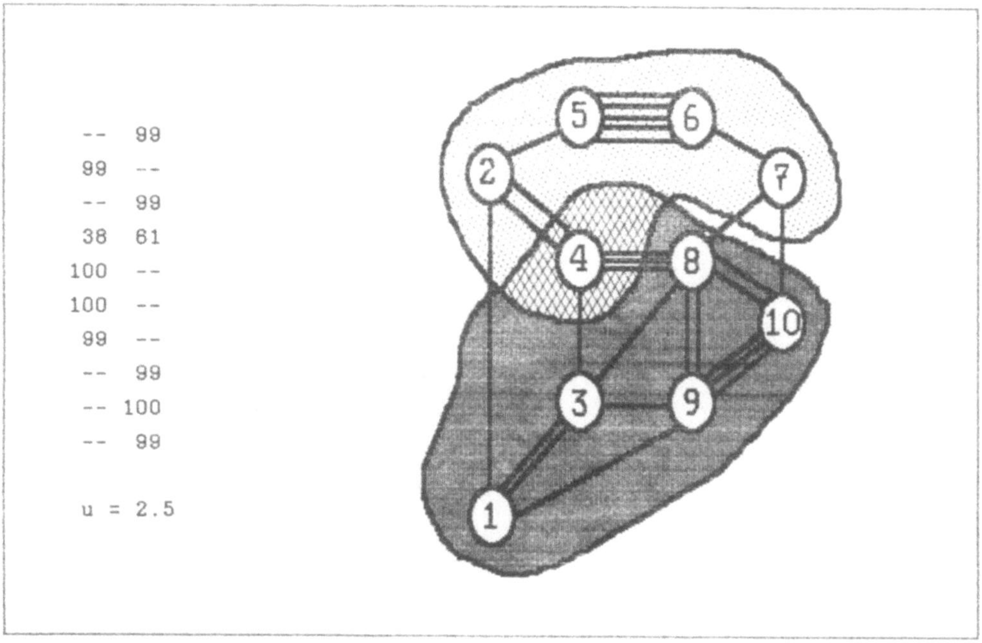
-- 99
99 --
-- 99
38 61
100 --
100 --
99 --
-- 99
-- 100
-- 99
u = 2.5
5
6
2
7
4
8
10
3
9
1

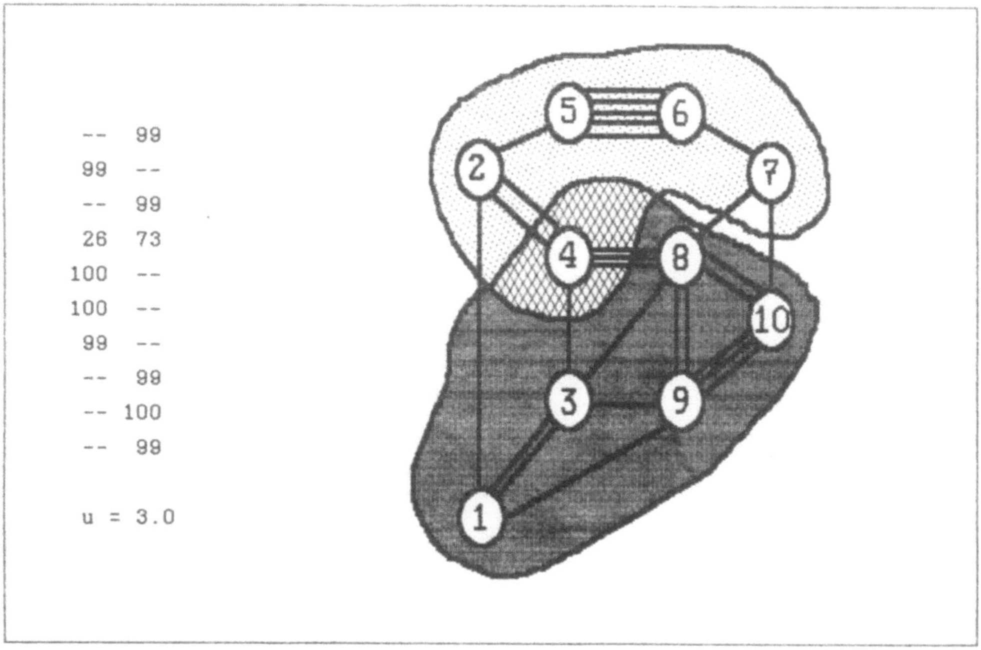
-- 99
99 --
-- 99
26 73
100 --
100 --
99 --
-- 99
-- 100
-- 99
u = 3.0
5
6
2
7
4
8
10
3
9
1

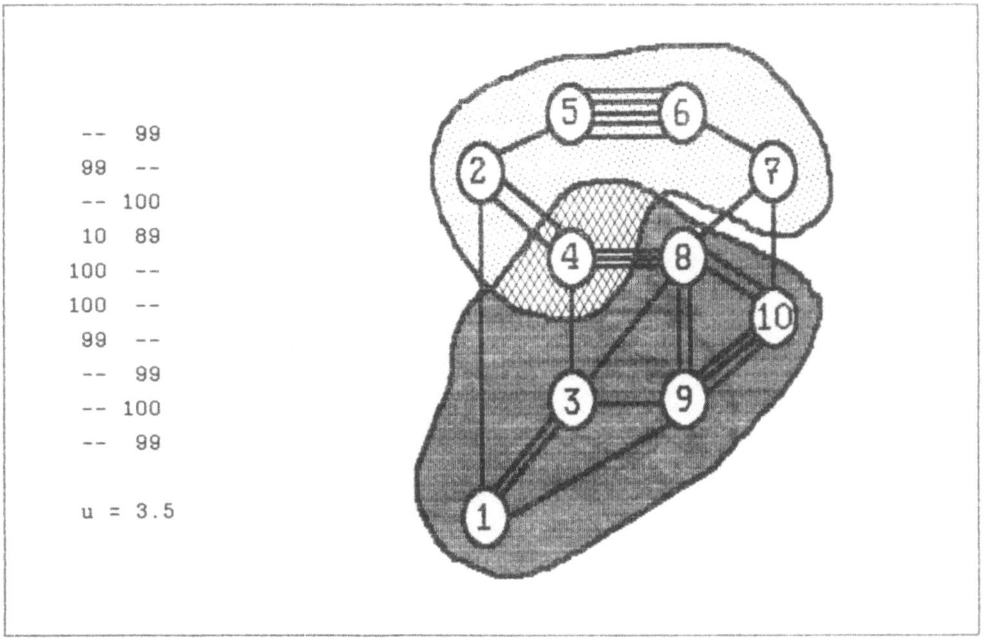
-- 99
99 --
-- 100
10 89
100 --
100 --
99 --
-- 99
-- 100
-- 99
u = 3.5
5
6
2
7
4
8
10
3
9
1

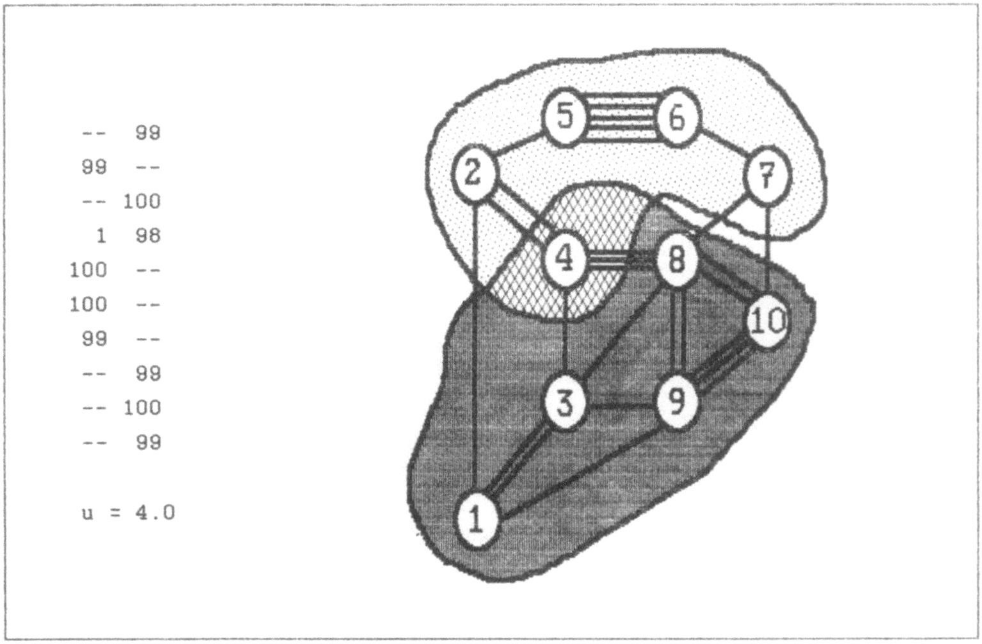
-- 99
99 --
-- 100
1 98
100 --
100 --
99 --
-- 99
-- 100
-- 99
u = 4.0
5
6
2
7
4
8
10
3
9
1

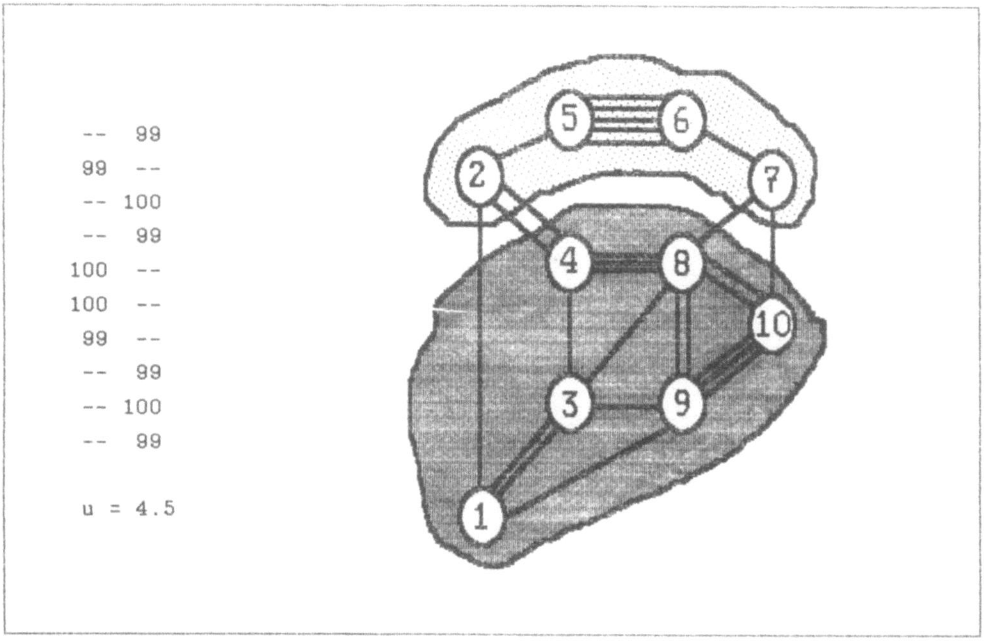

Figure 5.1.6 Assignment of vertices of the graph *G* to the partitions depending on the selection pressure *u* (*m* = 2)

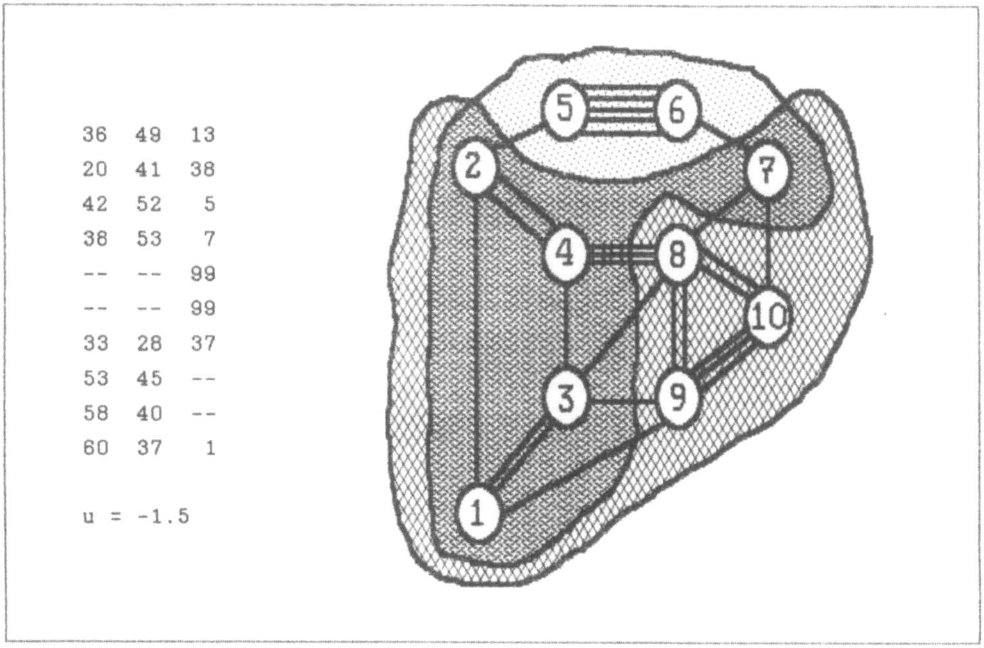

12	87	--
--	79	20
17	82	--
13	86	--
--	--	100
--	--	100
48	9	41
66	33	--
89	10	--
98	1	--

u = -1.0

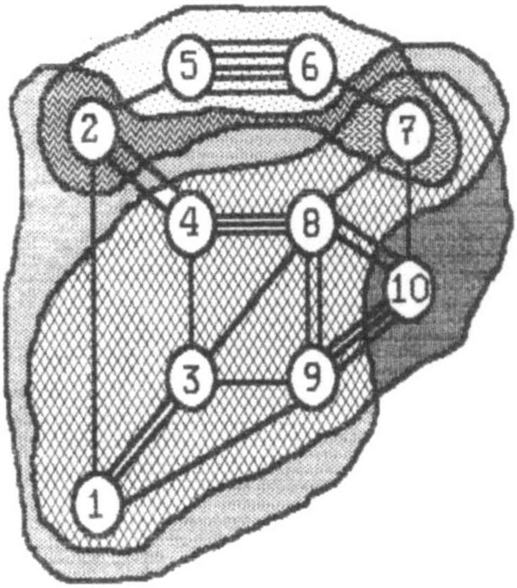

--	99	--
--	97	2
2	97	--
2	97	--
--	--	99
--	--	100
65	1	33
76	23	--
98	1	--
99	--	--

u = -0.5

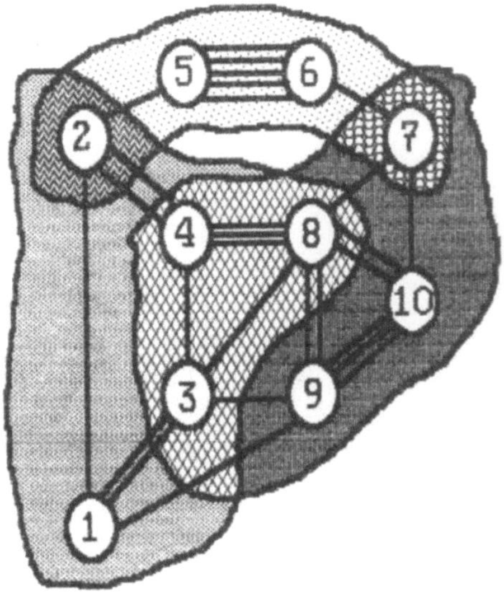

--	99	--
--	99	--
--	99	--
--	99	--
--	--	99
--	--	99
82	--	17
81	18	--
99	--	--
99	--	--

u = 0

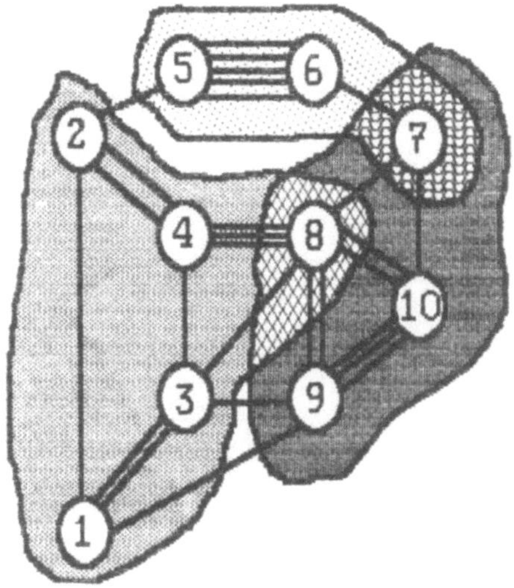

--	99	--
--	99	--
--	99	--
--	99	--
--	--	99
--	--	99
92	--	7
84	15	--
99	--	--
99	--	--

u = 0.5

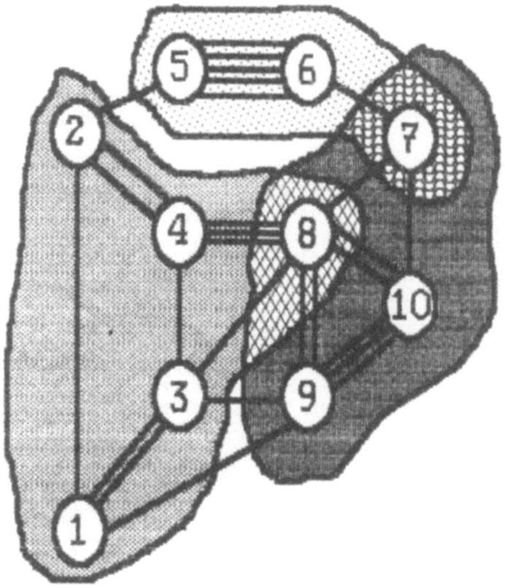

--	99	--
--	99	--
--	99	--
--	99	--
--	--	99
--	--	99
96	--	3
90	9	--
99	--	--
99	--	--

u = 1.0

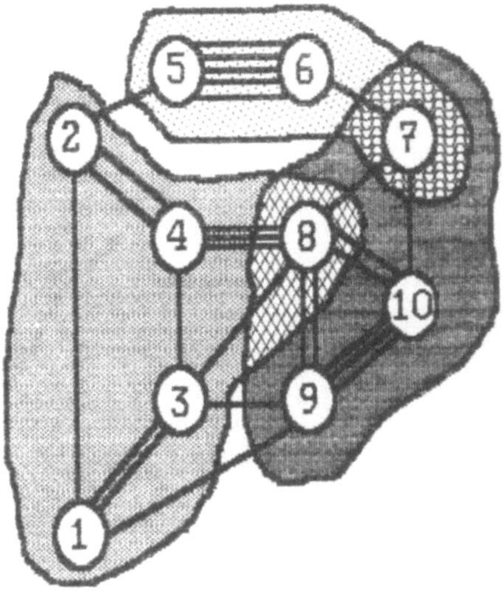

--	99	--
--	99	--
--	99	--
--	99	--
--	--	99
--	--	99
99	--	--
94	5	--
99	--	--
99	--	--

u = 1.5

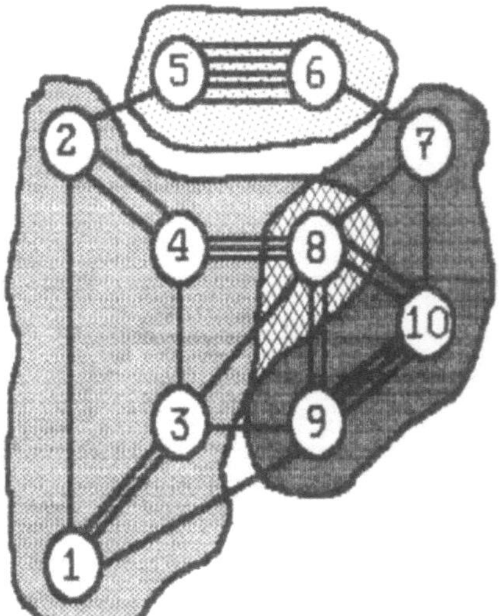

--	99	--
--	99	--
--	99	--
--	99	--
--	--	99
--	--	99
99	--	--
96	3	--
99	--	--
99	--	--

u = 2.0

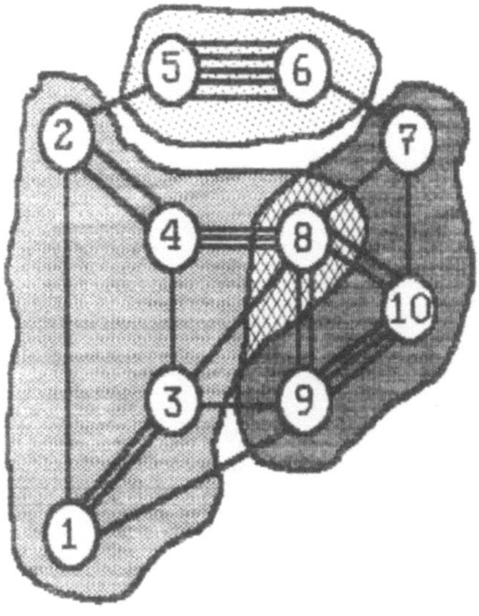

--	99	--
--	99	--
--	99	--
--	99	--
--	--	99
--	99	99
99	--	--
98	1	--
99	--	--
99	--	--

u = 2.5

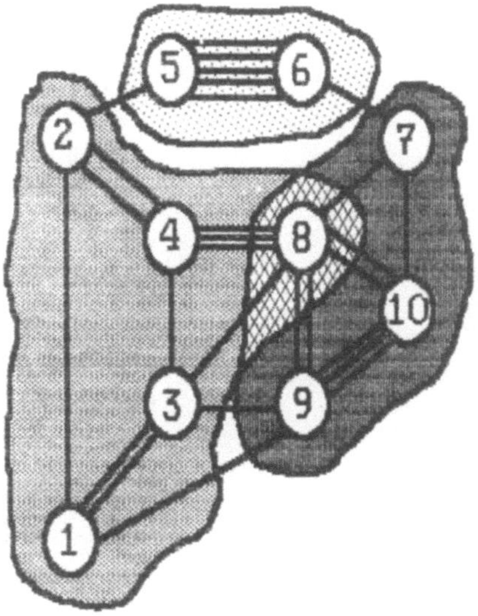

--	99	--
--	99	--
--	99	--
--	99	--
--	--	99
--	--	99
99	--	--
98	1	--
99	--	--
99	--	--

u = 3.0

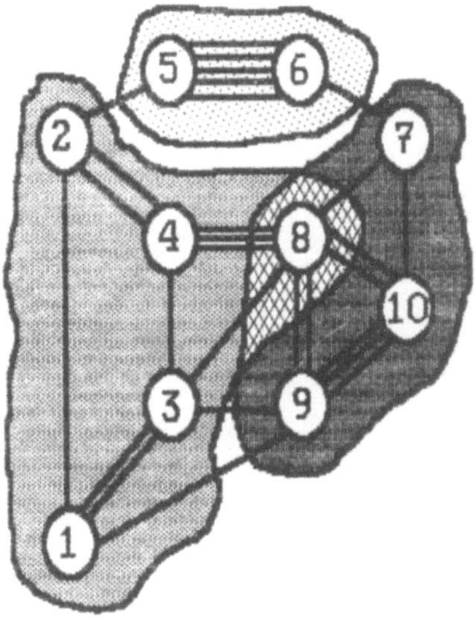

--	99	--
--	99	--
--	99	--
--	99	--
--	--	99
--	--	99
99	--	--
99	--	--
99	--	--
100	--	--

u = 3.5

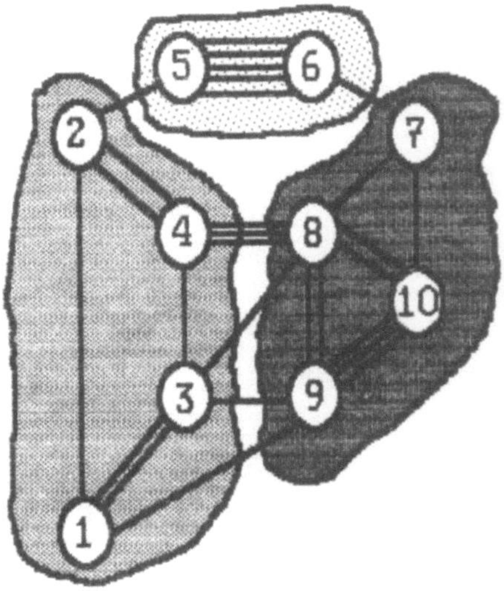

Figure 5.1.7 Assignment of vertices of the graph *G* to the partitions dependence on the selection pressure *u* (*m* = 3)

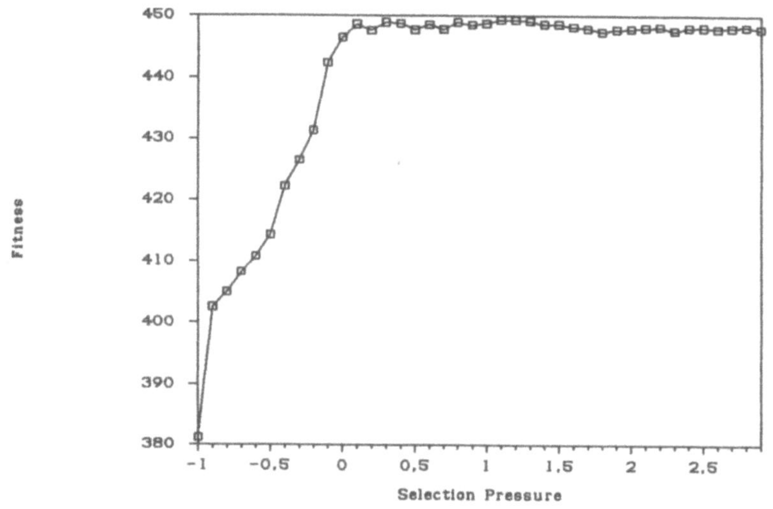

Figure 5.1.8 a) Fitness dependence on the selection pressure u **(**m **= 2)**

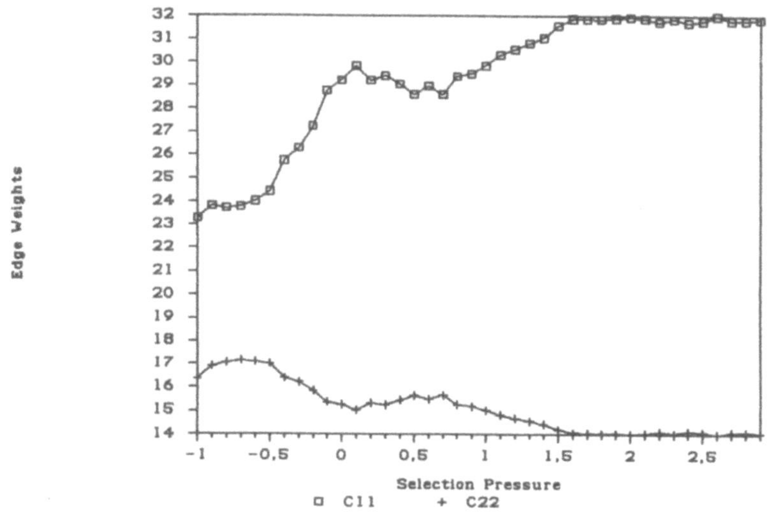

Figure 5.1.8 b) Edge weights C_{kk} **dependence on the selection pressure** u **(**m **= 2)**

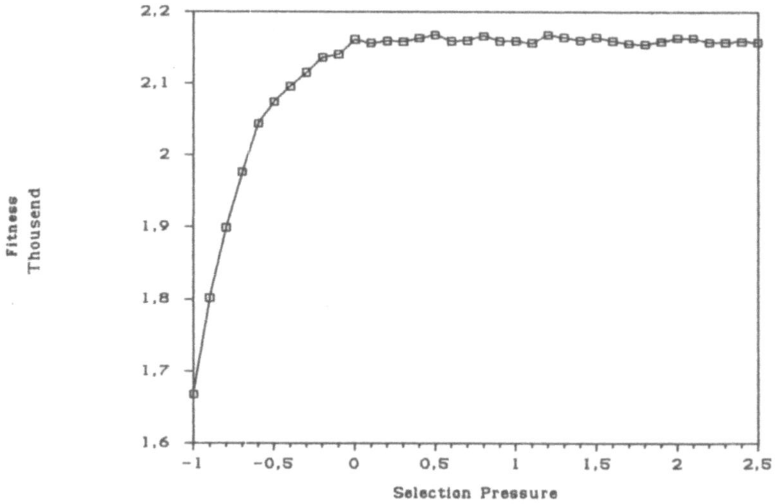

Figure 5.1.9 a) Fitness dependence on the selection pressure u (m = 3)

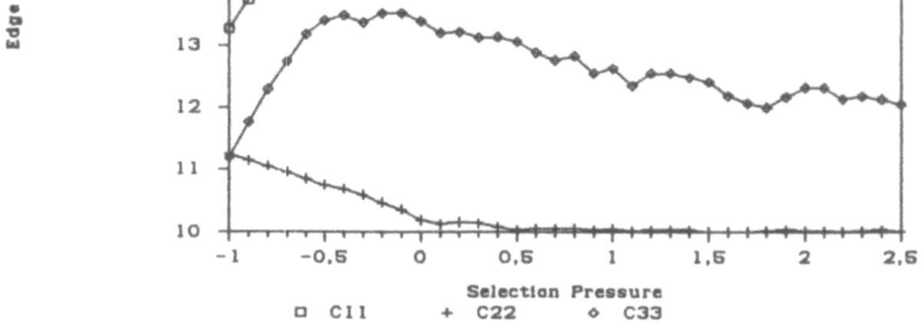

Figure 5.1.9 b) Edge weights C_{kk} dependence on the selection pressure u (m = 3)

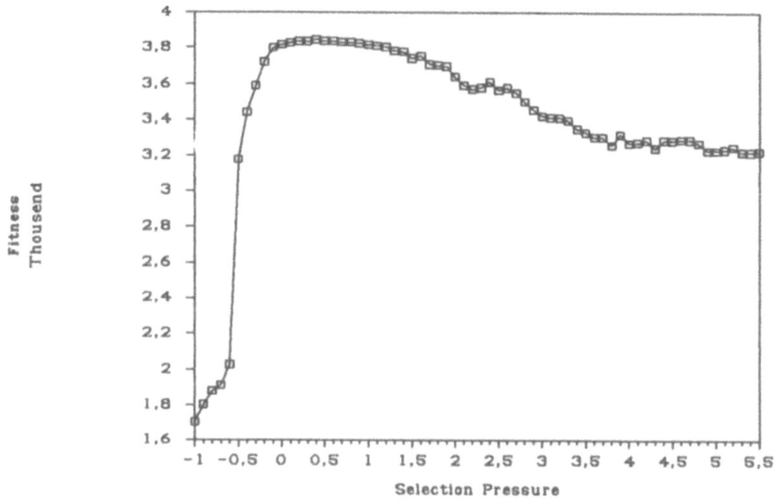

Figure 5.1.10 a) Fitness dependence on the selection pressure u (m = 4)

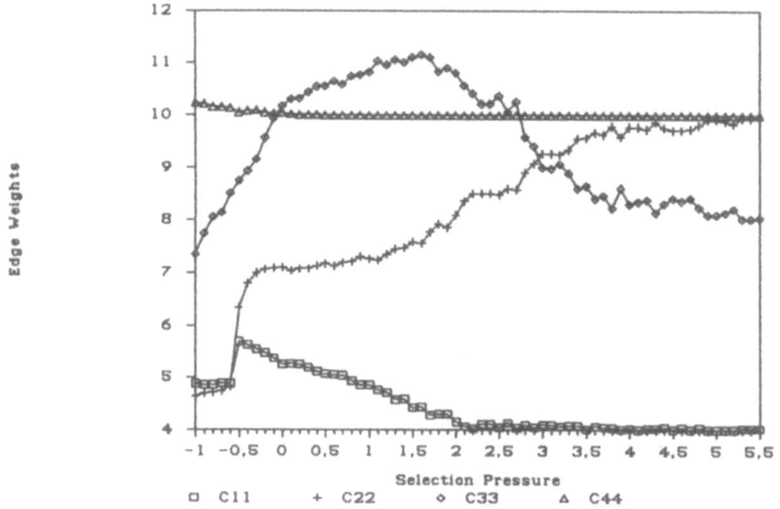

Figure 5.1.10 b) Edge values C_{kk} dependence on the selection pressure u (m = 4)

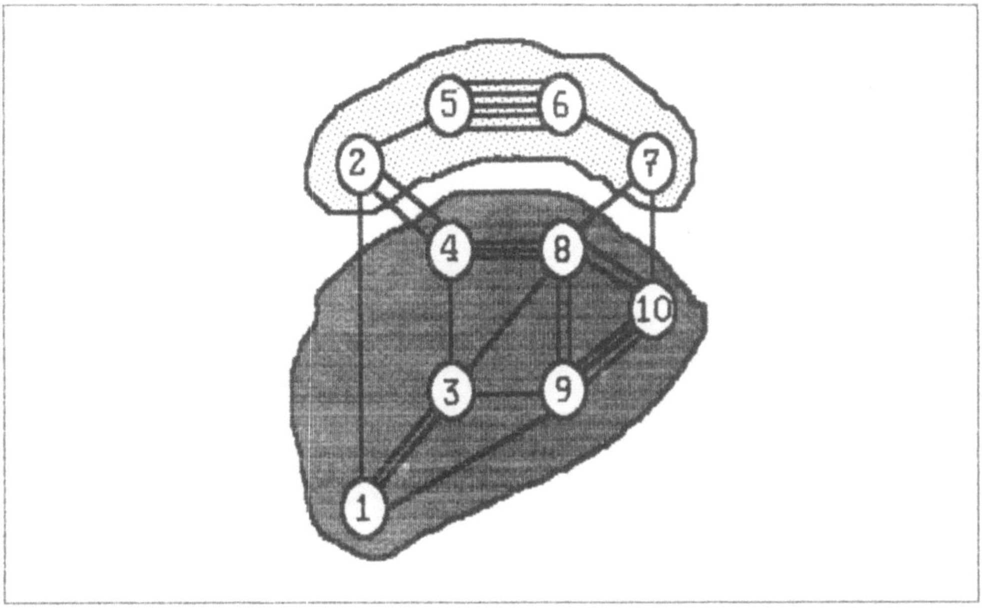
5
6
2
7
4
8
10
3
9
1

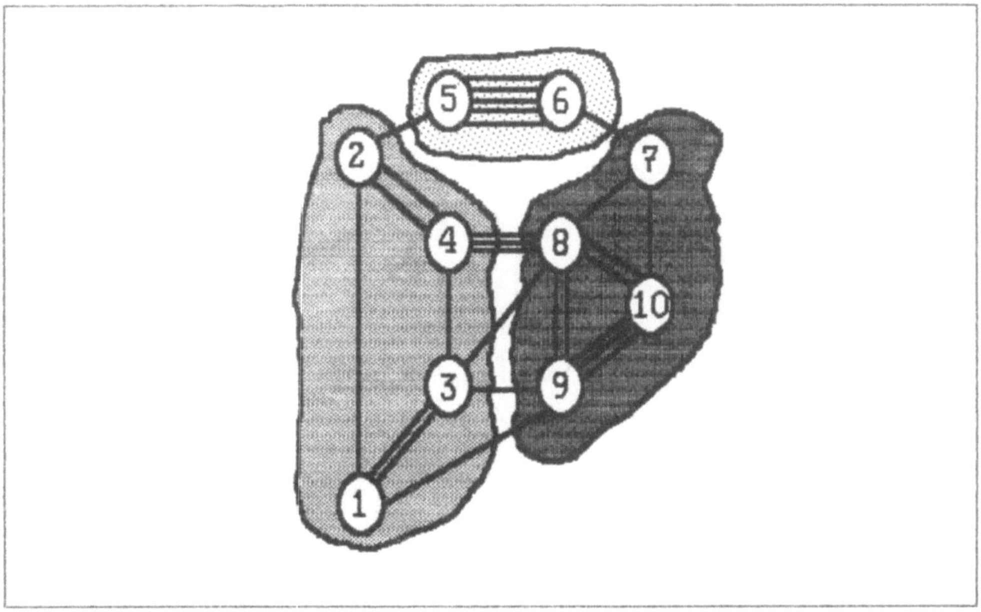
5
6
2
7
4
8
10
3
9
1

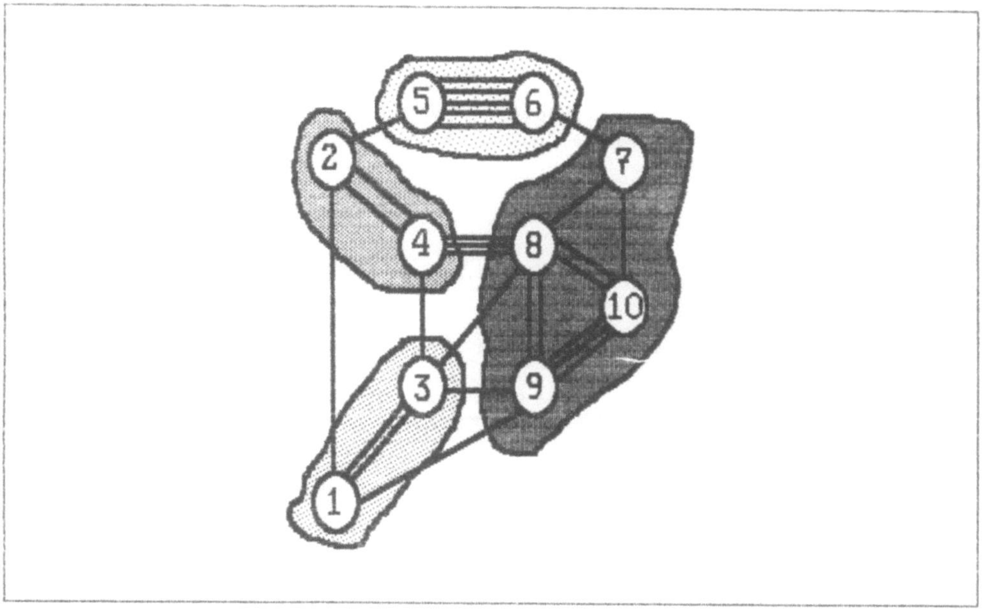

Figure 5.1.11 Decomposition of the graph G into different partitions (m = 2 , 3, and 4)

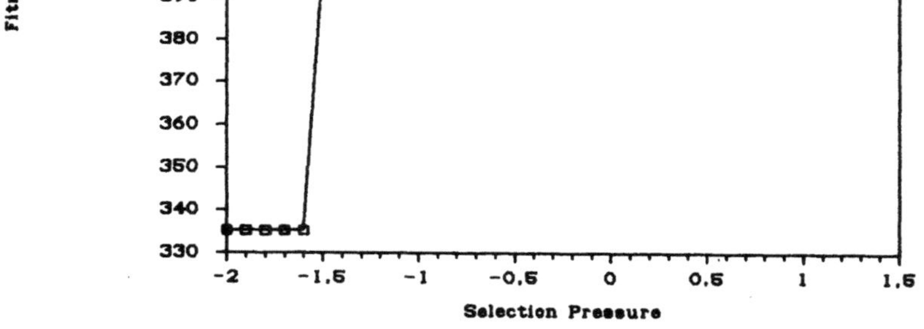

Figure 5.1.12 a) Fitness dependence on the selection pressure u (e = 0.0001, h = 0.1, a = 0, m = 2)

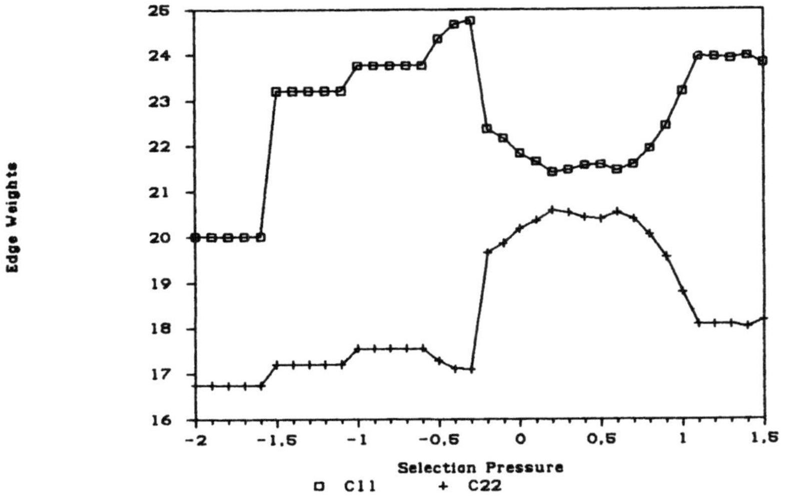

Figure 5.1.12 b) Edge values C_{kk} dependence on the selection pressure u (e = 0.0001, h = 0.1, a = 0, m = 2)

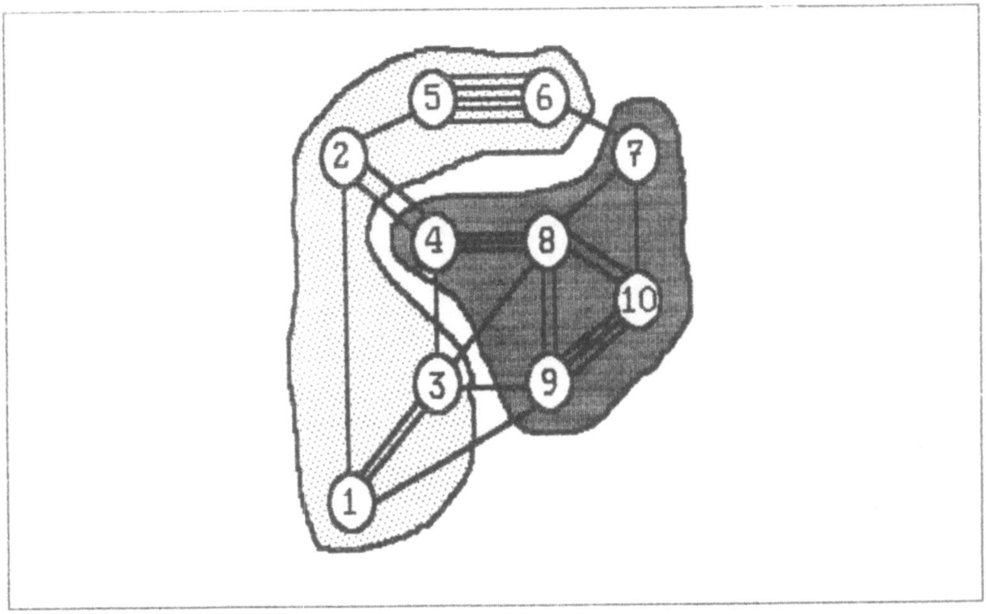

Figure 5.1.12 c) Assignments of vertices of the graph G to the partitions (e = 0.0001, h = 0.1, a = 0, m = 2)

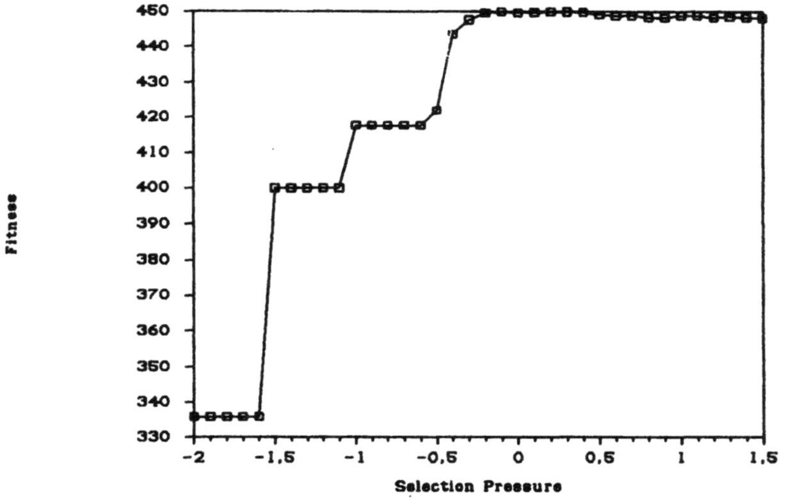

Figure 5.1.13 a) Fitness dependence on the selection pressure u
(e = 0.0001, h = 0.1, a = 2, m = 2)

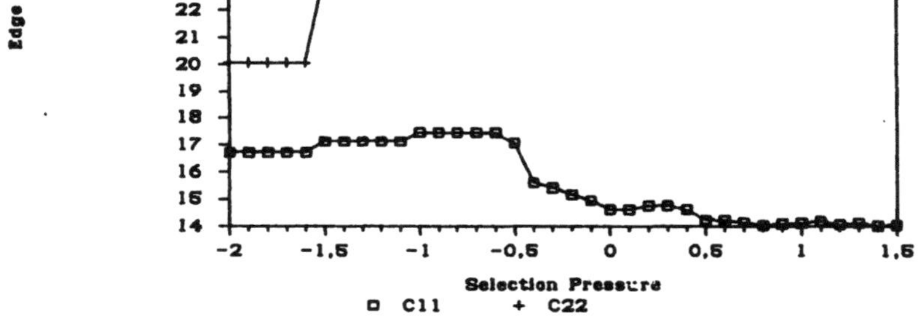

Figure 5.1.13 b) Edge values C_{kk} dependence on the selection pressure u
(e = 0.0001, h = 0.1, a = 2, m = 2)

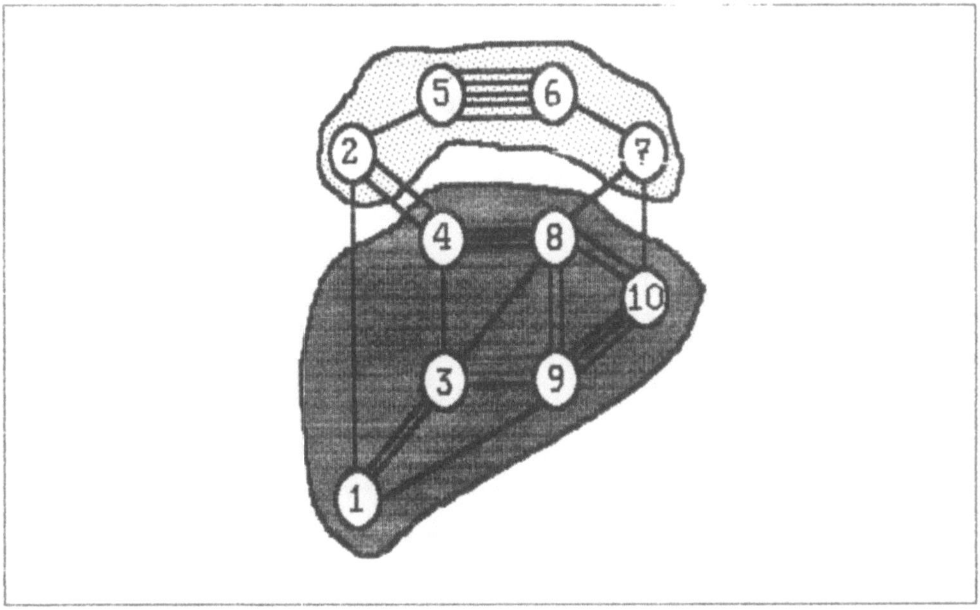

Figure 5.1.13 c) Assignment of vertices of the graph G to the partitions (e = 0.0001, h = 0.1, a = 2, m = 2)

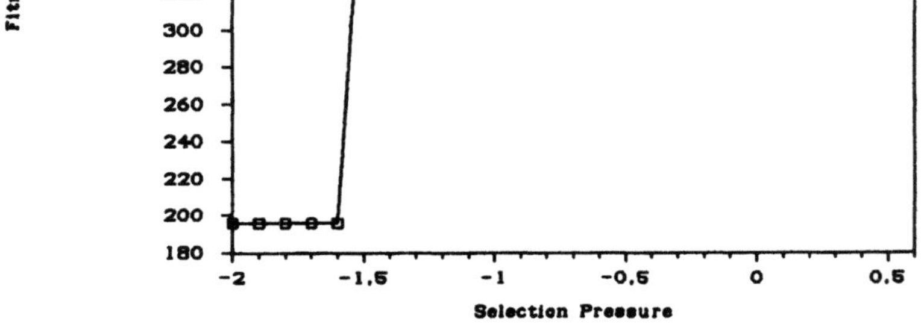

Figure 5.1.14 a) Fitness dependence on the selection pressure u (e = 0.0001, h = 0.1, a = 6, m = 2)

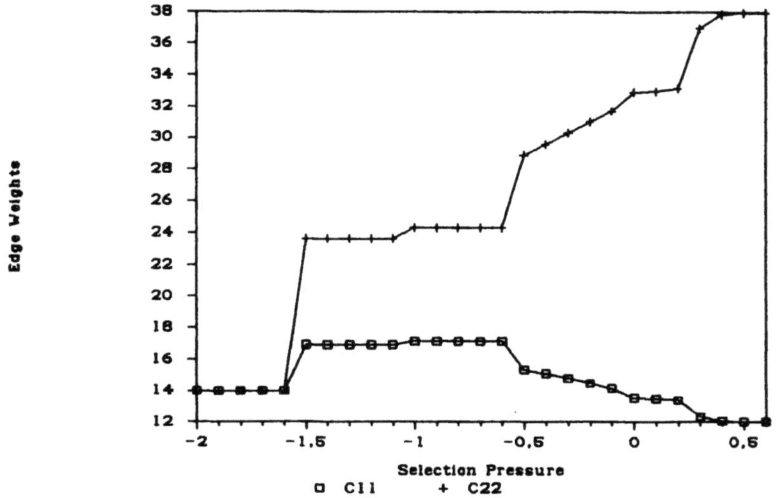

Figure 5.1.14 b) Edge values C_{kk} dependence on the selection pressure u (e = 0.0001, h = 0.1, a = 6, m = 2)

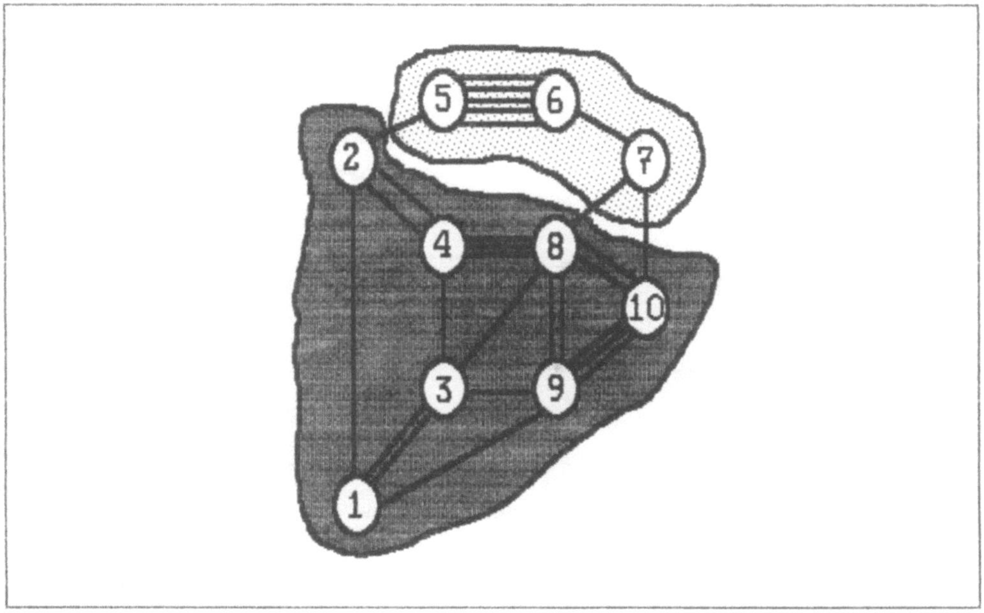

Figure 5.1.14 c) Assignment of vertices of the graph G to the partitions (e = 0.0001, h = 0.1, a = 6, m = 2)

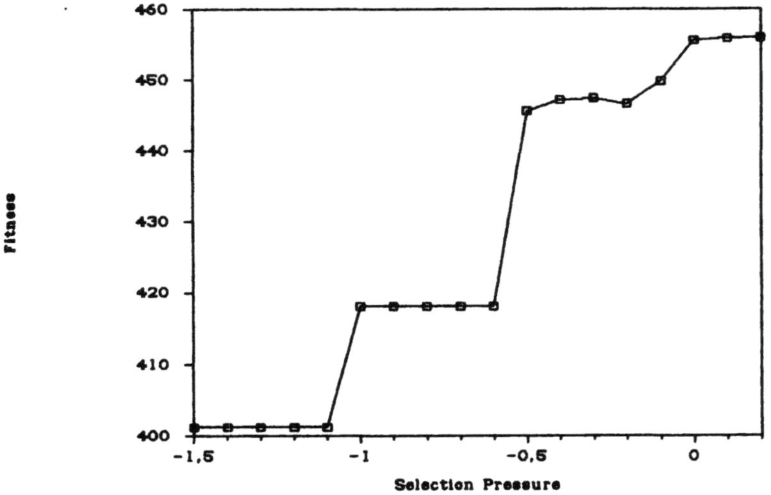

Figure 5.1.15 a) Fitness dependence on the selection pressure *u* (*e* = 0.0001, *h* = 0.1, *a* = 10, *m* = 2)

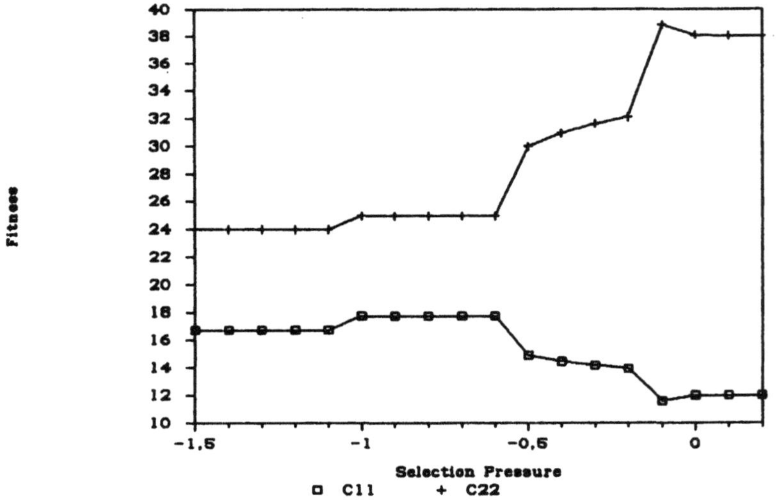

Figure 5.1.15 b) Edge values C_{kk} dependence on the selection pressure *u* (*e* = 0.0001, *h* = 0.1, *a* = 10, *m* = 2)

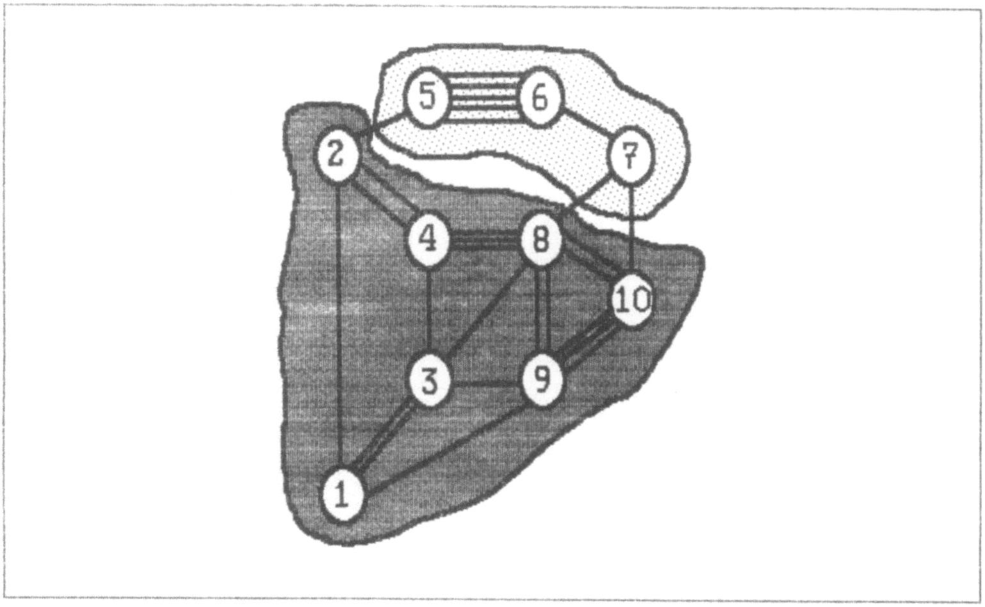

Figure 5.1.15 c) Assignment of vertices of the graph G to the partitions (e = 0.0001, h = 0.1, a = 10, m = 2)

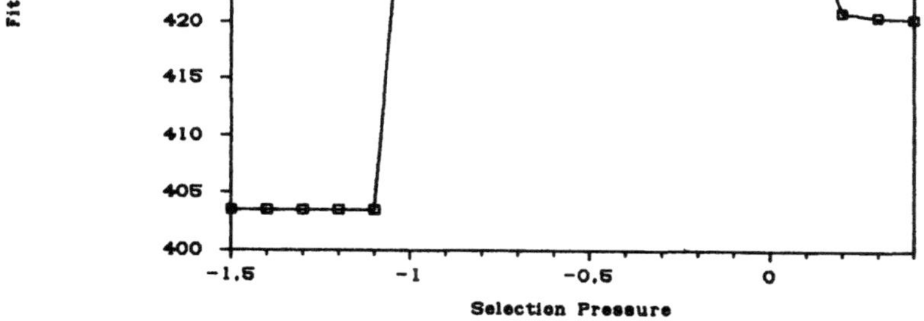

Figure 5.1.16 a) Fitness dependence on the selection pressure u (e = 0.0001, h = 0.1, a = 20, m = 2)

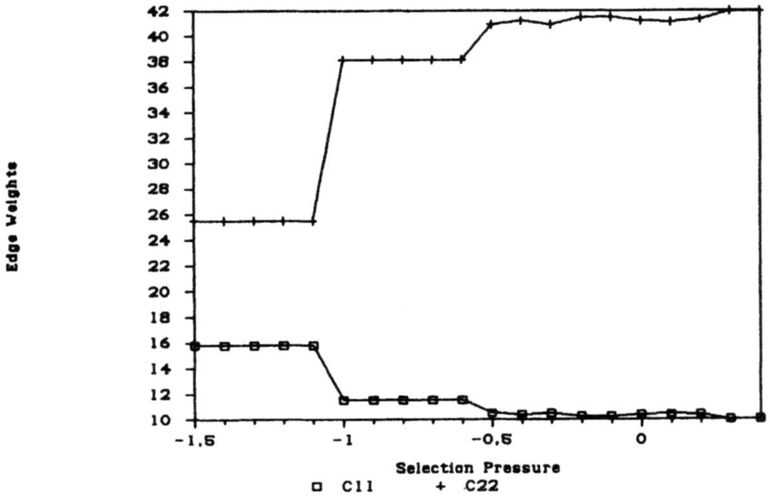

Figure 5.1.16 b) Edge values C_{kk} dependence on the selection pressure u (e = 0.0001, h = 0.1, a = 20, m = 2)

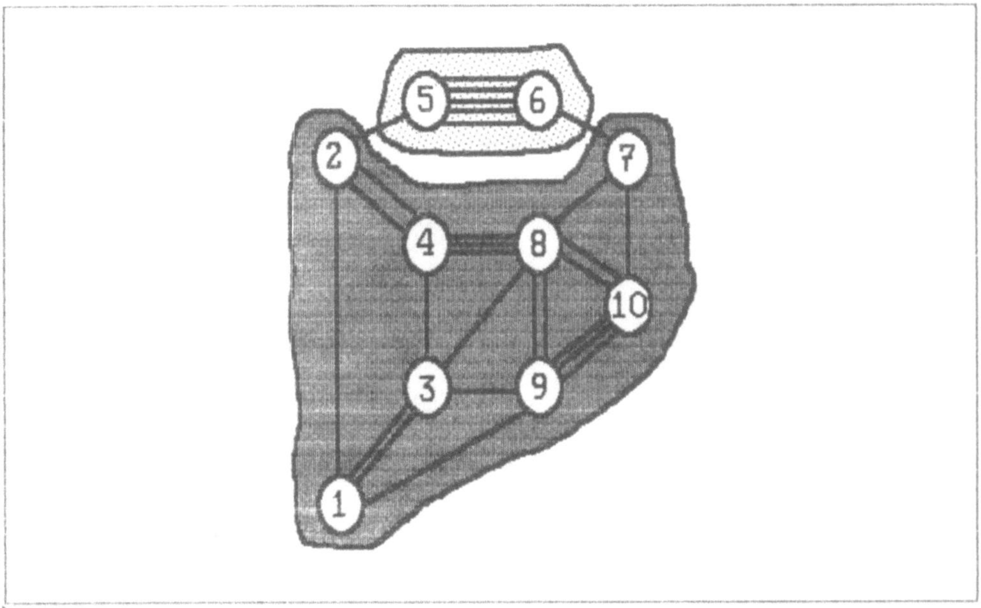

Figure 5.1.16 c) Assignment of vertices of the graph G to the partitions (e = 0.0001, h = 0.1, a = 20, m = 2)

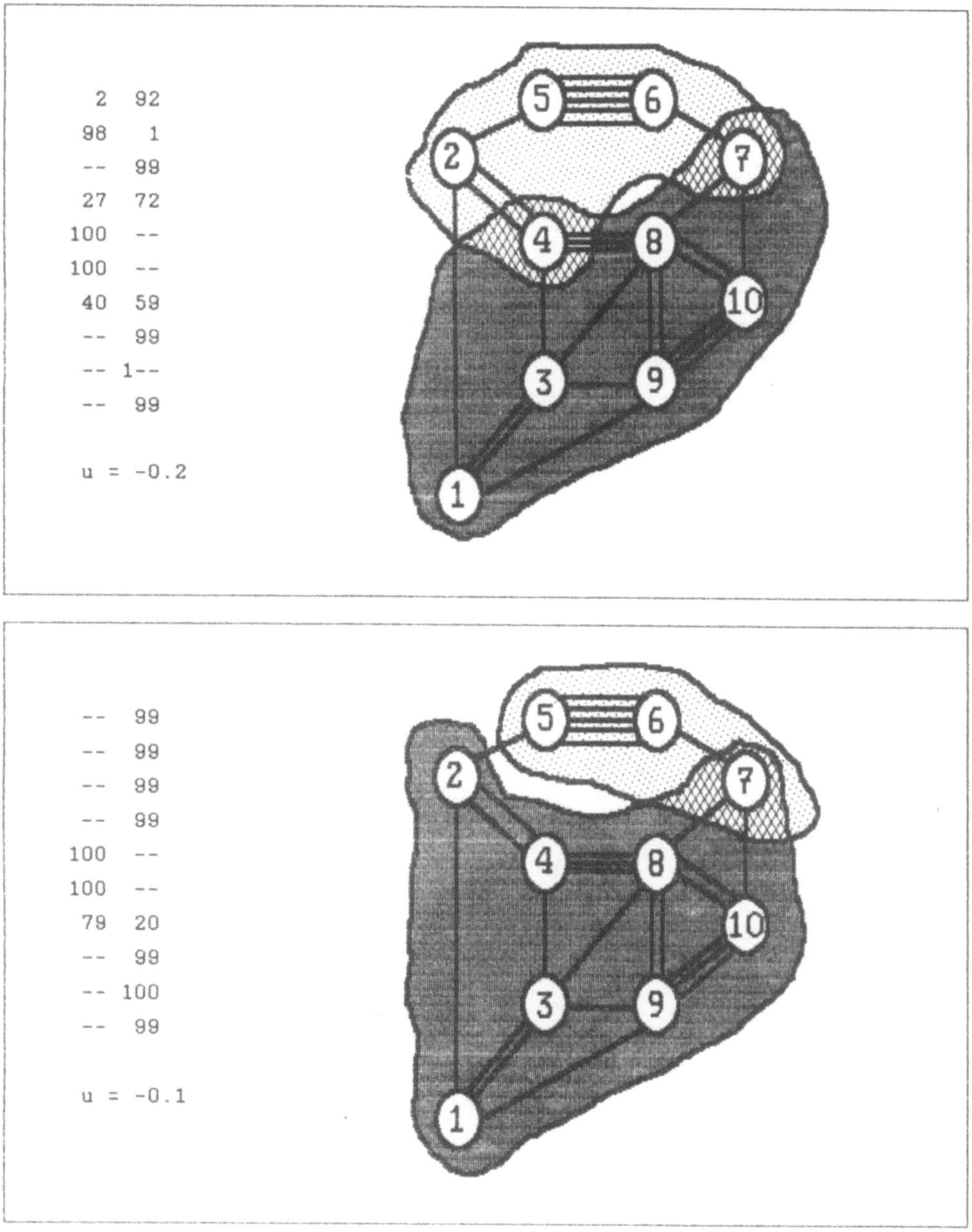

Figure 5.1.17 "Switching" of vertex 2 of the graph G from partition 1 to partition 2 (e = 0.0001, h = 0.1, a = 10, m = 2)

balance control	edge value partition 1	edge value partition 2	cuts between partitions
0	9	12	7
2	7	16	5
6	6	19	3
10	6	19	3
20	5	21	2

Table 5.1.1 Edge values C_{kk} and cuts between the partitions for different values of the balance control *a*

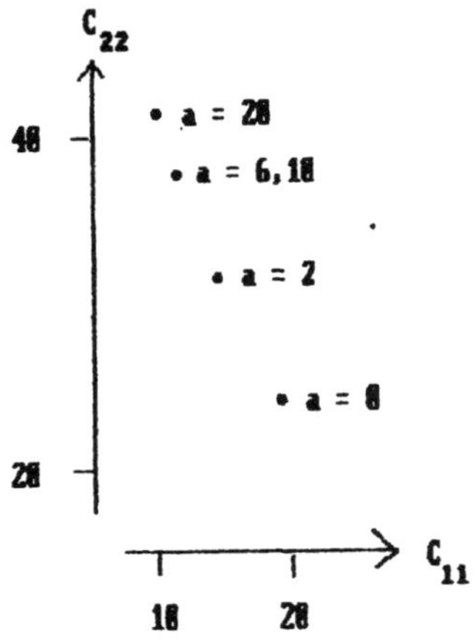

Figure 5.1.18 Efficient solutions for different values of the balance control *a*

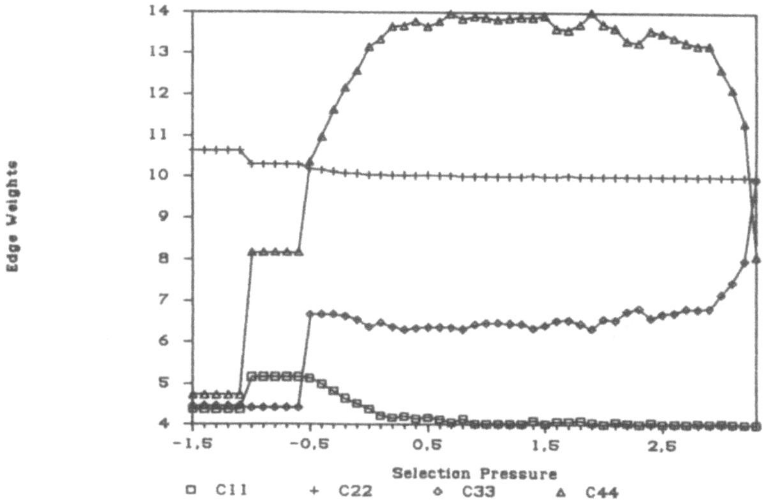

Figure 5.1.19 a) Edge values C_{kk} dependence on the selection pressure u (e = 0.001, h = 0.1, a = 5, m = 4)

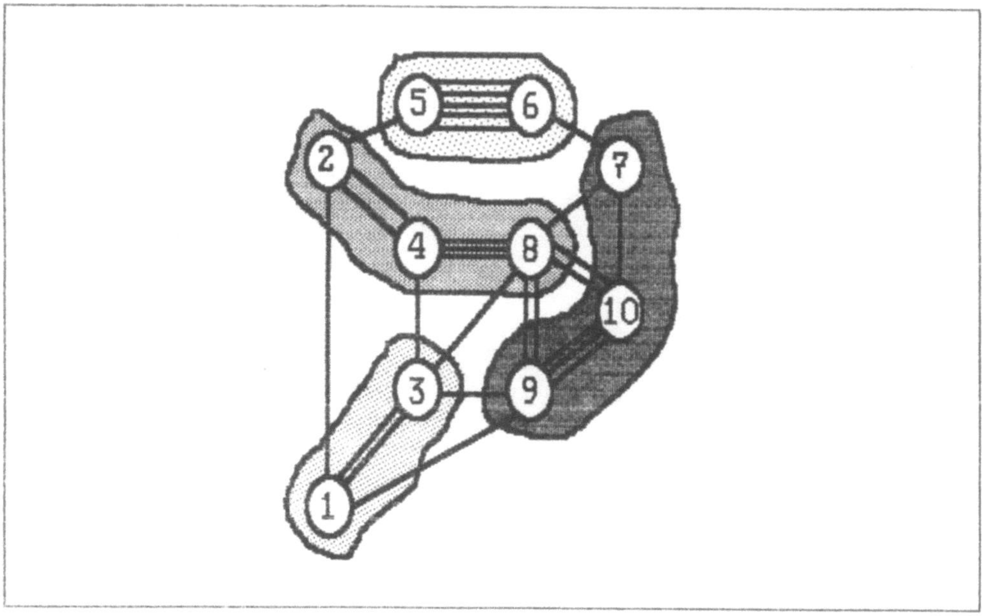

Figure 5.1.19 b) Assignment of vertices of the graph G to the partitions (e = 0.001, h = 0.1, a = 5, m = 4)

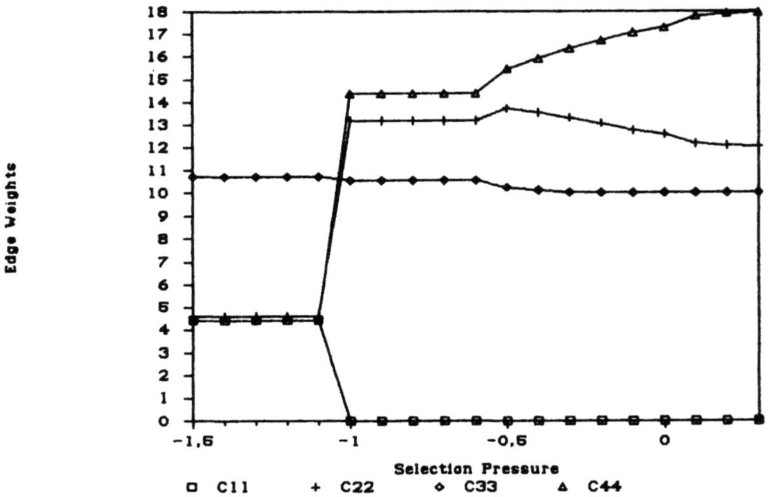

Figure 5.1.20 a) Edge values C_{kk} dependence on the selection pressure u (e = 0.001, h = 0.1, a = 10, m = 4)

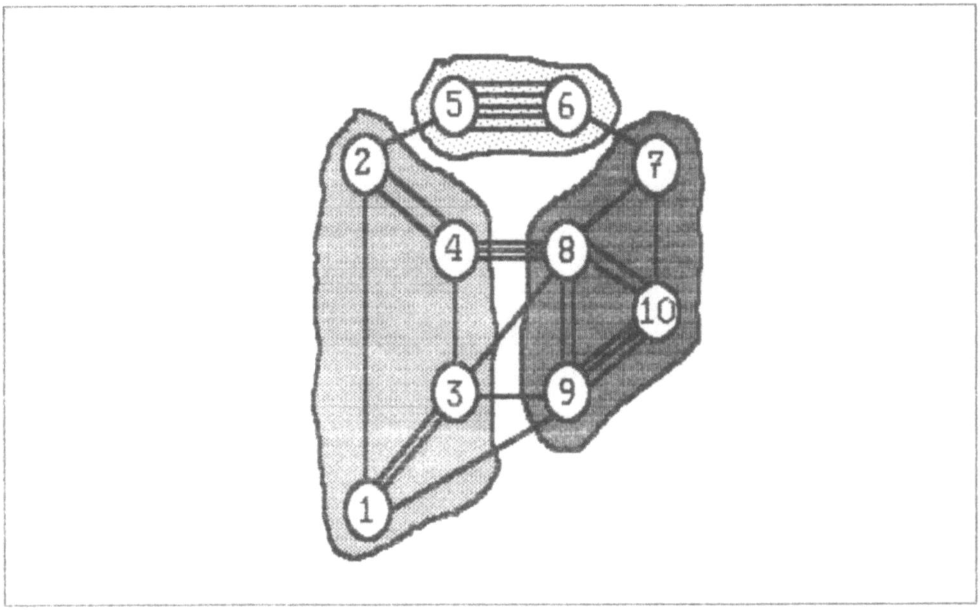

Figure 5.1.20 b) Assignment of vertices of the graph G to the partitions (e = 0.001, h = 0.1, a = 10, m = 4)

5.1.3. Comparison of Methods

We want to compare now some self-organization methods for the graph partitioning problem be means of a somewhat more representative example **B41A** which was delivered by the Institut für Nachrichtentechnik. The comparison includes three different approaches for which we will draw some conclusions concerning the efficiency and the solution performance.

The first algorithm **KOLOSS** rests on models from statistical mechanics which were first used for the solution of combinatorial optimization problems by Kirkpatrick, Gelatt & Vecchi (1983) and by Cerny (1983). It was implemented by von Trzebiatowski (1985) from the Sektion Mathematik of the Humboldt-Universität at Berlin.

The second algorithm **UTA** is based on a fuzzy "thermodynamic" approach which was developed by Richardt (1985) from the Institut für Nachrichtentechnik at Berlin

The third algorithm **PNS** results from the preceding population genetic models which were developed at the Forschungsbereich Mathematik und Informatik of the Akademie der Wissenschaften der DDR (Voigt (1985)).

Here we give a brief description of the approaches. Detailed explanations by the corresponding authors are contained in Voigt (Ed.) (1985).

The essence of the thermodynamic approach consists in the following (Kirkpatrick, Gelatt & Vecchi (1983)): We consider an ensemble of identical systems (Landau & Lifschitz (1979)). Every configuration within this ensemble is characterized by the set of atom positions $\{r_i\}$ of the system. There is assumed only weak interaction between the atoms. The probability of such a configuration is given by the Boltzmann distribution

$$x_i = C \exp\left(\frac{-E(\{r_i\})}{k_B T}\right).$$

Here $E(\{r_i\})$ is the configuration energy, k_B is the Boltzmann constant, and T is the temperature. C is a normalization term

Let us assume that we start an annealing process with a high initial temperature. If we lower the temperature sufficiently slow then we have at last as a ground state of the system a configuration with global minimal energy $E(\{r_i\})$.

If we consider $E(\{r_i\})$ as an optimization functional and the set $\{r_i\}$ as the set of optimization parameters, this can be used as an optimization algorithm. It is very essential that the optimization process is controlled by the control parameter T in a sufficiently smooth way. A too fast annealing, i.e. a quenching, leads to degenerated ground states, i.e. to local minima.

For the implementation of this approach the algorithm of Metropolis & Teller (1953) is used. This is a Monte-Carlo method for the effective simulation of a collection of atoms at equilibrium for a given temperature.

Starting with an arbitrary configuration the position of an atom will be disturbed slightly by random. If we have for the resulting energy change a nonpositive value this new configuration will be accepted as a starting configuration for a next position change. If it is greater zero the case will be treated statistically. Let y be a random variable which is uniformly distributed in (0,1). If we have now

$$y < \exp\left(\frac{-\delta E}{k_B T}\right)$$

then the new configuration will be also accepted. In all other cases the old configuration will be retained. By this simulation the system evolves corresponding to a Boltzmann distribution. It is interesting that during the simulation process increases of the optimization functional may occur.

Here it should be noted once more: Essential for a global optimization is a slow cooling, i.e. a slow decrease of the temperature.

The essential of the fuzzy "thermodynamic" approach (Richardt (1985)) consists in a direct computation of the probabilities x_i by an iterative algorithm. The aim consists in the optimization of partial goals. The optimization process will also be controlled by one or more global parameters. It is remarkable that Richardt (1980, 1983) detected the influence of such global parameters to cluster algorithms very early and independent of analogies to thermodynamic models. For the control of such cluster algorithms he formulated the "principle of minimal intensification".

For the population genetic approach of the preceding sections similar propositions hold concerning the selection pressure. To reach a "global" optimum a sufficiently slow increase of the selection pressure is necessary.

The comparison will bring only rough estimates of the efficiency and the solution performance of the proposed methods because different computer types and different programming languages were used.

The results are gathered up in Voigt, von Trzebiatowski & Richardt (1985). Here we want to consider one of the test examples in detail.

The net list of example **B41A** is presented in Table 5.1.2. Figure 5.1.21 shows the corresponding graph.

As one performance measure we used the total number of cuts between the partitions for different number of partitions m = 2, 4 , 5. For this first attempt we renounced further performance measures which could play a role corresponding to Section 5.1.1.

As a second performance measure we used the number of iterations because a comparison of the computer time was impossible. The used computers and programming languages were too different. For the different methods "iteration" has a different meaning:

KOLOSS

Iteration: Interchange of two vertices of G from different partitions and computation of the functional

UTA and **PNS**

Iteration: Computation of a new matrix $[x'_{ik}]$ from a matrix $[x_{ik}]$

The results (total number of cuts, number of iterations), the corresponding partition sizes (number of vertices of G within the partitions), and the control schedule of the algorithm are gathered up in Table 5.1.3. The assignment of vertices to the partitions are contained in Table 5.1.4. Figures 5.1.22 to 5.1.30 show selected assignments to the partitions.

Data set = B41A

Fac	Wire						
	01 =	1	2				
	02 =	1	3				
	03 =	4	27				
	04 =	1	2	3	4	2"	38
	05 =	8	41				
2	06 =	11	41				
	07 =	11	18	23	40		
	08 =	18	23	41			
3	09 =	25	38				
	10 =	25	31				
2	11 =	25	36				
3	12 =	26	38				
2	13 =	28	36				
	14 =	25	28	38			
2	15 =	5	12				
2	16 =	5	29				
2	17 =	6	29				
	18 =	6.	12				
3	19 =	7	30				
4	20 =	9	29				
	21 =	9	30				
	22 =	4	5	6	7		
	23 =	5	6	7	28		
	24 =	5	6	12			
	25 =	6	7	12			
3	26 =	15	35				
	27 =	15	20				
5	28 =	16	21				
	29 =	16	35				
2	30 =	19	35				
	31 =	19	22				
2	32 =	20	35				
2	33 =	20	22				
	34 =	15	16	20			
2	35 =	15	16	19			
	36 =	16	21	33			
4	37 =	10	32				
2	38 =	14	32				
	39 =	10	14	37			
	40 =	10	14	17	37		
	41 =	10	14	17			
3	42 =	13	24				
	43 =	24	39				
2	44 =	24	34				
	45 =	24	34	39			

Net list

Fac	Wire					
	46 =	8	13			
	47 =	8	22			
	48 =	8	21			
	49 =	9	36			
	50 =	10	31			
	51 =	11	30			
3	52 =	12	22			
	53 =	12	24			
	54 =	12	19			
	55 =	13	37			
	56 =	14	33			
	57 =	17	36			
	58 =	17	30			
	59 =	17	38			
	60 =	17	33			
	61 =	18	24			
	62 =	19	34			
	63 =	22	33			
	64 =	22	40			
	65 =	23	39			
	66 =	26	31			
2	67 =	29	41			
	68 =	31	34			
	69 =	33	34			
	70 =	34	35			
	71 =	34	37			
	72 =	34	36			
	73 =	2	10	31		
	74 =	3	26	38		
	75 =	4	26	27		
	76 =	8	11	13	17	39
	77 =	8	11	39		
	78 =	8	17	18	23	40
	79 =	9	33	34		
	80 =	11	13	40		
	81 =	17	31	34	37	
	82 =	17	29	30		
	83 =	17	21	30	39	
	84 =	28	35	33		
	85 =	28	33	37		

Table 5.1.2 Net list of example **B41A**

partitions	method	partitions size	total cuts	object assignments
2	KOLOSS	19/22	22	P1:5-9/11/12/15/16/19-23/29/30/35/40/41 P2:1-4/10/13/14/17/18/24-28/31-34/36-39
		20/21	20	P1:1-4/10/13/14/17/24-28/31-34/36-38 P2:5-9/11/12/15/16/18-23/29/30/35/39-41
	UTA	18/23	18	P1:1-4/10/14/17/25-28/31-34/36-38 P2:5-9/11-13/15/16/18-24/29/30/35/39-41
	PNS	20/21	20	P1:1-4/10/13/14/17/24-28/31-34/36-39 P2:5-9/11/12/15/16/18-23/29/30/35/40/41
		19/22	21	P1:4-7/9/12/15/16/19-22/27-30/33/35/36 P2:1-3/8/10/11/13/14/17/18/23-26/31/32/34
4	KOLOSS	10/11/10/10	47	P1:4-7/9/11/12/29/30/41 P2:10/14/17/25/26/28/31/32/36-38 P3:8/15/16/19-22/33/35/40 P4:1-3/13/18/23/24/27/34/39
	UTA	8/11/10/12	35	P1:12/15/16/19/20-22/35 P2:10/13/14/17/24/31-34/37/39 P3:1-4/25-28/36/38 P4:5-9/11/18/23/29/30/40/41
	PNS	11/11/11/8	38	P1:8/11/13/18/23/24/34/37/39-41 P2:1-4/25-28/31/36/38 P3:5-7/9/10/12/14/17/29/30/32 P4:15/16/19-22/33/35
5	KOLOSS	9/9/7/9/7	44	P1:15/19/20/22/28/33-36 P2:5-7/9/12/16/21/29/30 P3:1/3/4/25-27/38 P4:8/11/13/18/23/24/39/41 P5:2/10/14/17/31/32/37
	UTA	10/7/8/7/9	38	P1:10/14/17/28/31-34/36/37 P2:5-7/9/12/29/30 P3:1-4/25-27/38 P4:15/16/19-22/35 P5:8/11/13/18/23/24/39-41
	PNS	9/9/7/9/7	39	P1:8/11/13/18/23/24/39-41 P2:1-4/25-27/36/38 P3:5-7/9/12/29/30 P4:10/14/17/28/31-34/37 P5:15/16/19-22/35

Table 5.1.5 Assignments for example **B41A**

partitions	method	partition size	cut potentials	total cuts	control parameters	iterations
2	KOLOSS	19/22	21	21	To=5	40807
		20/21	20	20		30733
	UTA	18/23	18	18	EV=1.5,3.0	37
	PNS	20/21	20	20	u=-1.0,-0.9, -0.8,-0.5, -0.2	172
		19/22	21	21	u=-1.5,-1.0, -0.5	114
4	KOLOSS	10/11/10/10	36	47	To=4	44747
	UTA	8/11/10/12	35	37	EV=1.3,1.5, 3.0	81
	PNS	11/11/11/8		38	u=-1.0,-0.5, -0.3,-0.2, -0.1	229
5	KOLOSS	9/9/7/9/7	42	44	To=5	42046
		9/9/7/9/7	40	47	To=4	41076
	UTA	10/7/8/7/9	36	38	EV=3.0	22
	PNS	9/9/9/7/7		39	u=-4.0,-3.0, -2.0,-1.0, -0.5	165

Table 5.1.3 Results for example **B41A**

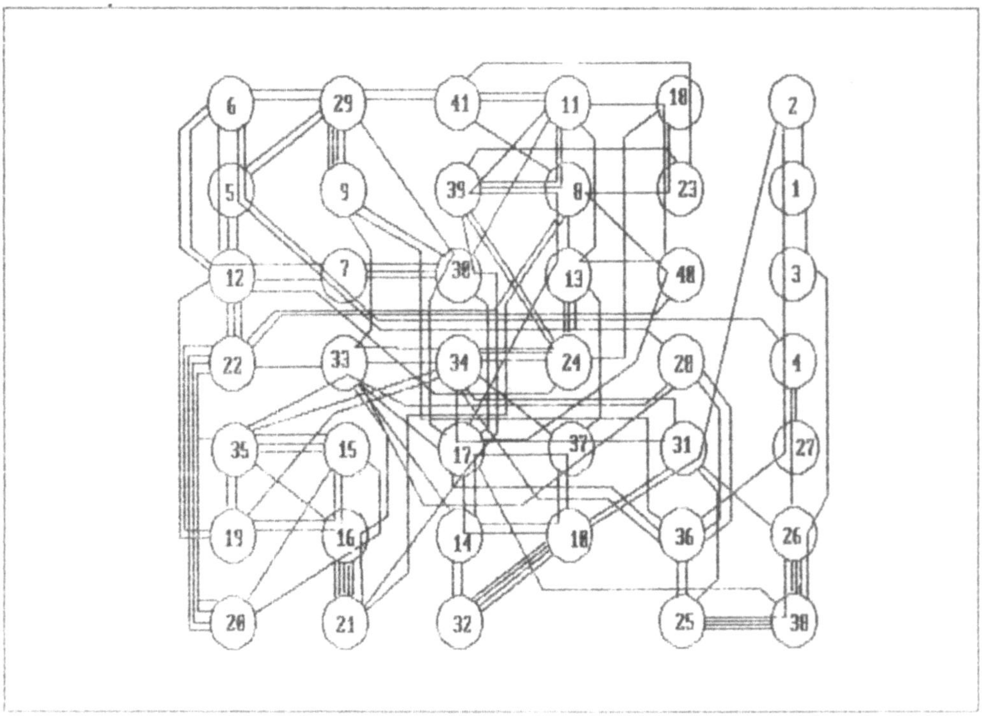

Figure 5.1.21 Graph of example **B41A**

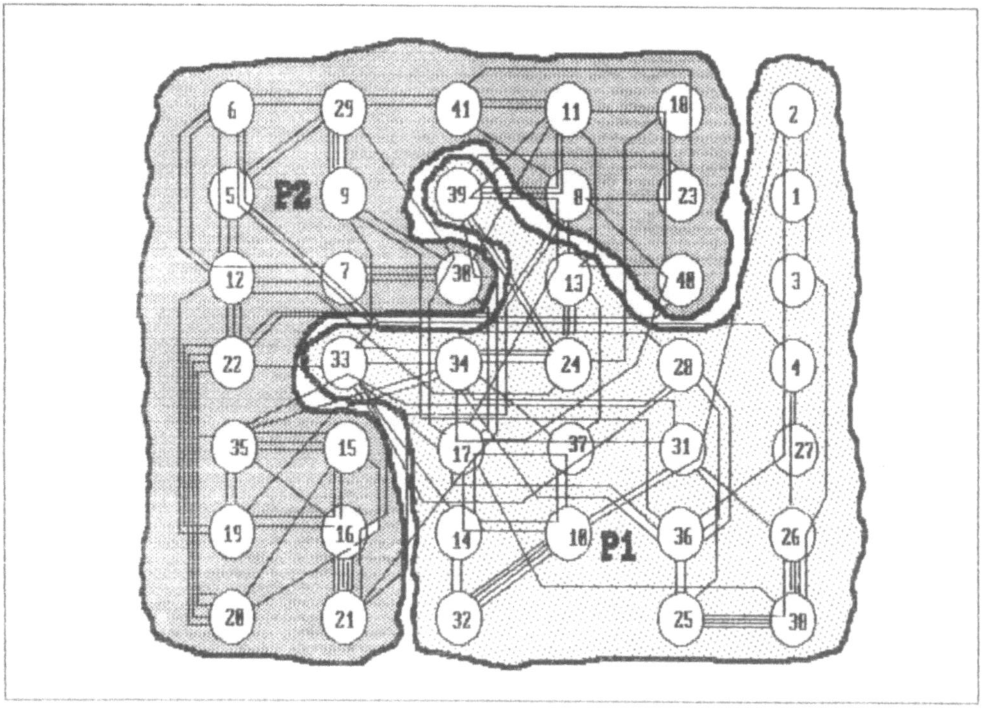

Figure 5.1.22 **B41A**, **PNS**, 20 cuts, partition sizes 20/21

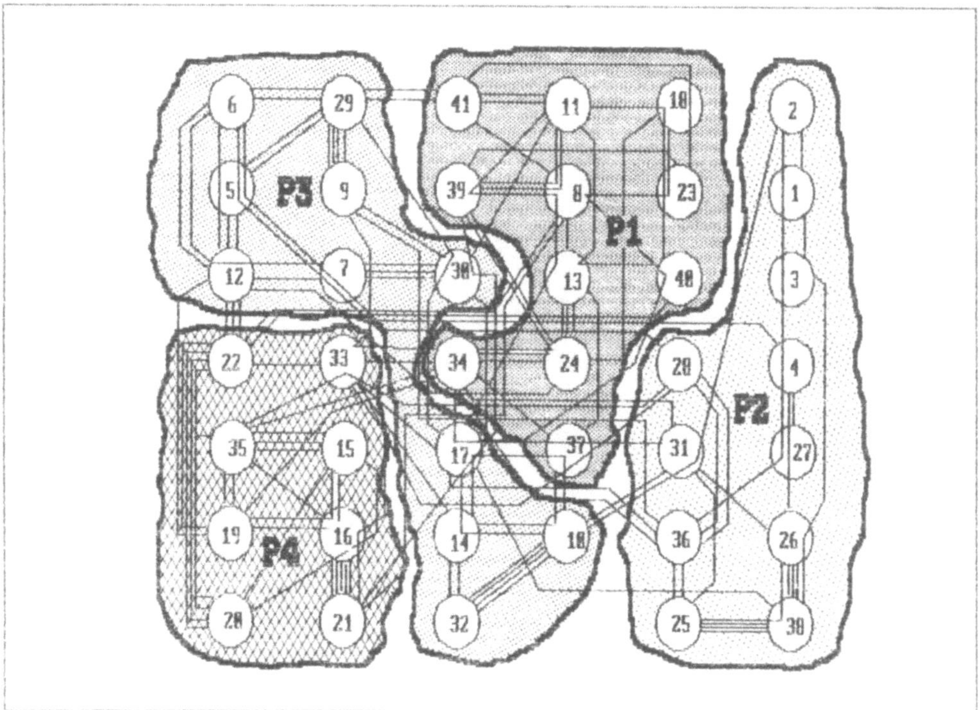

Figure 5.1.23 **B41A**, **PNS**, 38 cuts, partition sizes 11/11/11/8

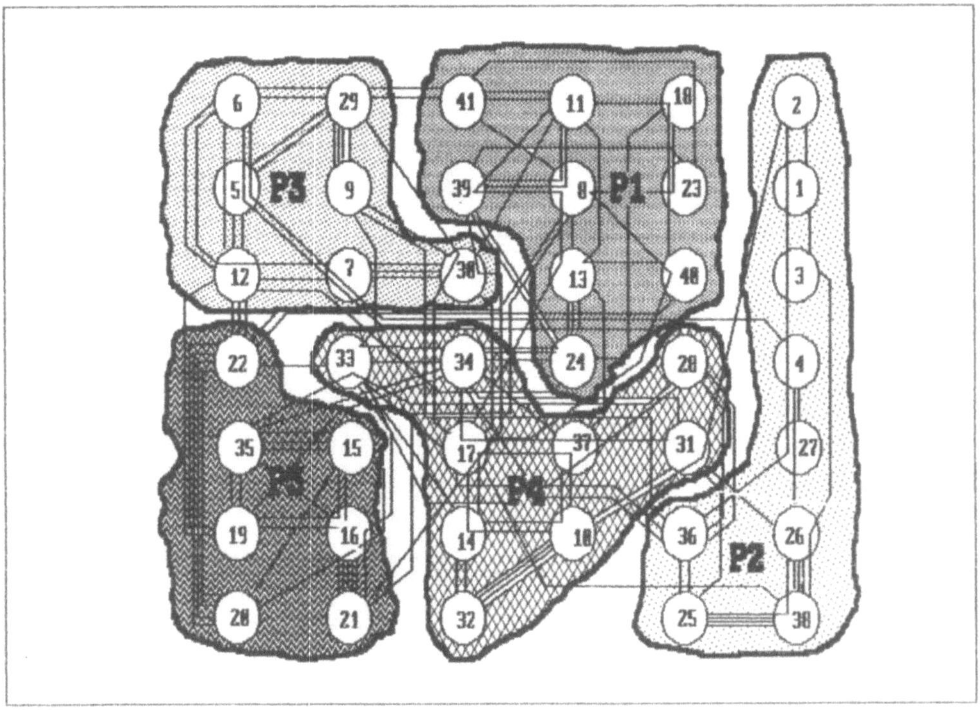

Figure 5.1.24 **B41A, PNS**, 39 cuts, partition sizes 9/9/9/7/7

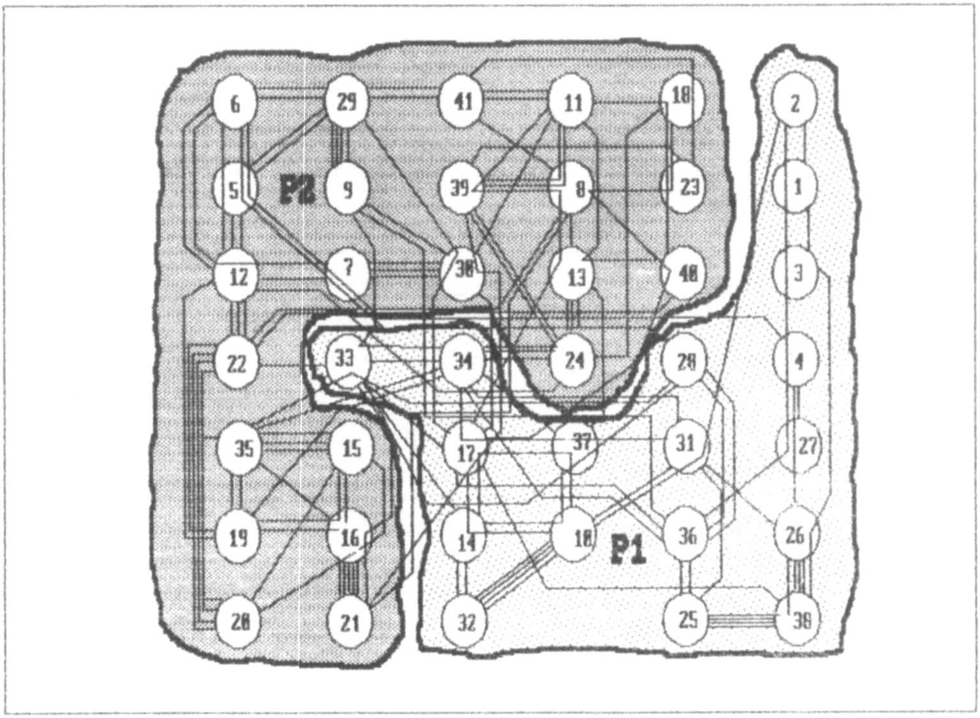

Figure 5.1.25 **B41A, UTA**, 18 cuts, partition sizes 18/23

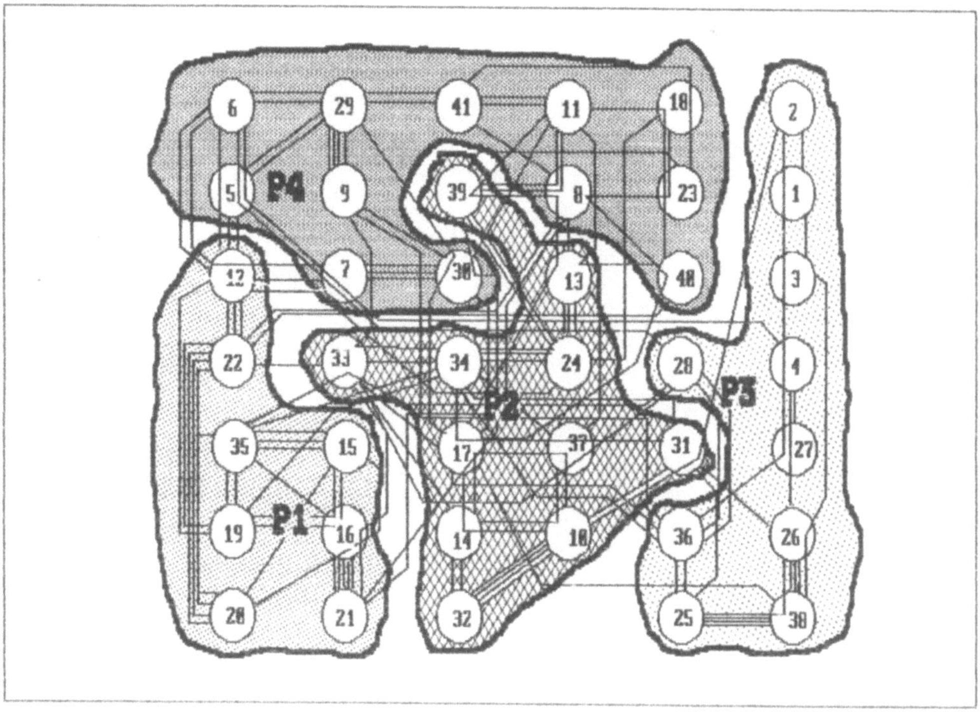

Figure 5.1.26 **B41A, UTA**, 35 cuts, partition sizes 12/11/10/8

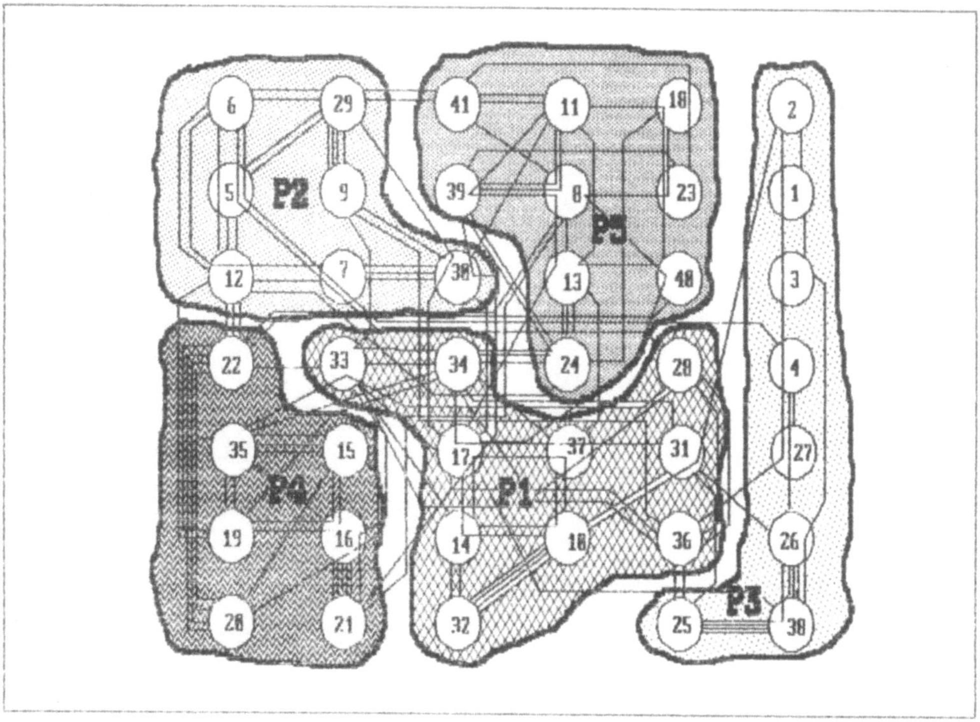

Figure 5.1.27 **B41A, UTA**, 38 cuts, partition sizes 10/9/8/7/7

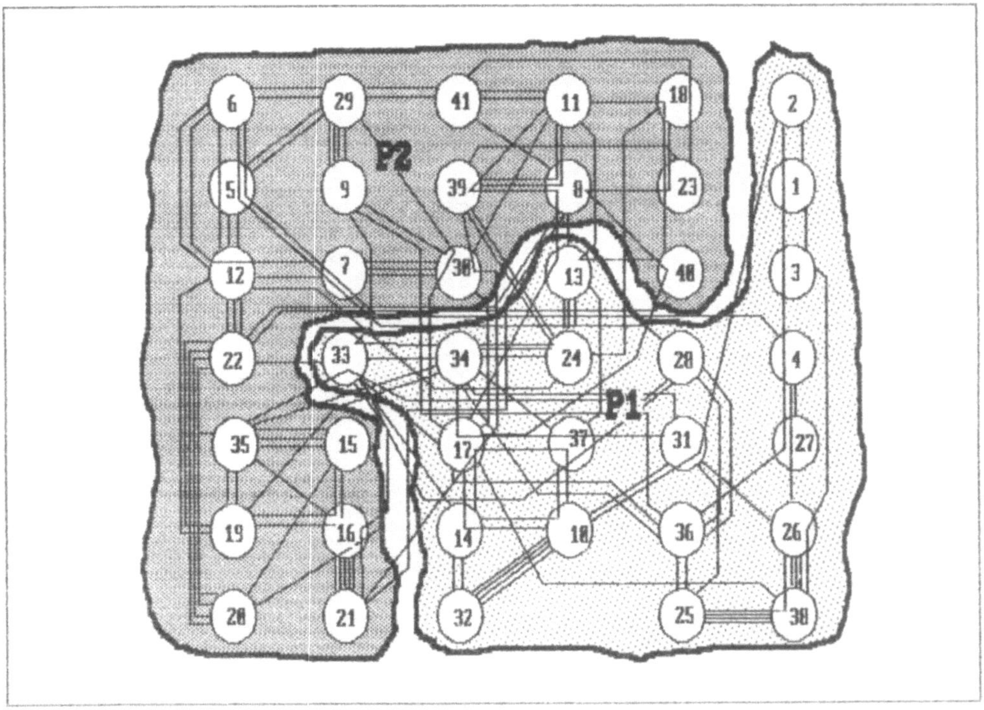

Figure 5.1.28 **B41A, KOLOSS**, 20 cuts, partition sizes 20/21

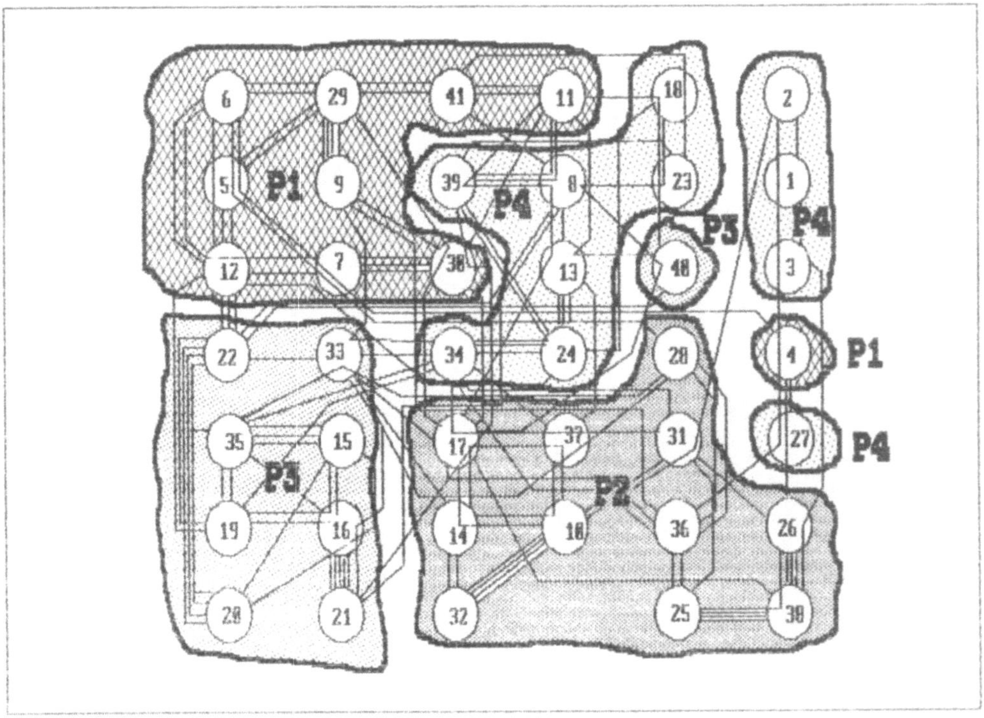

Figure 5.1.29 **B41A, KOLOSS**, 47 cuts, partition sizes 11/10/10/10

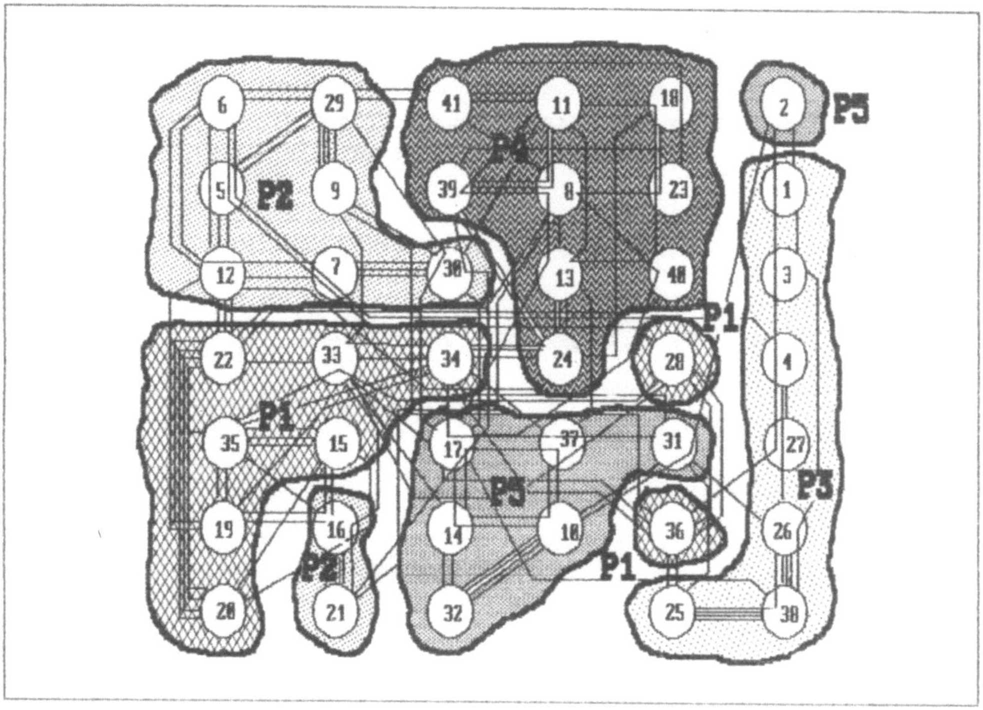

Figure 5.1.30 **B41A**, **KOLOSS**, 44 cuts, partition sizes 9/9/9/7/7

5.2. The Module Placement Problem

Module placement plays an important role in the design of VLSI circuits with respect to wirability and wire length minimization. On the other hand it should be mentioned that the problem of site allocation with respect to minimal transportation efforts and the task allocation problem e.g. for computer systems can be formalized in much the same way.

5.2.1. Fitness

In case of the module placement problem the potential function can be derived in the following way (Voigt & Santibanez-Koref (1988)).

Let us assume that there are given modules i, $i = 1,\ldots,n$. The modules are connected by wires where G_{ij} is the number of wires between module i and module j. The modules should be placed on a regular grid with k rows and l columns. The total length of the wires for a placement of modules to grid points has to be minimized according to a Manhattan distance.

This can be formalized as a maximization problem with

$$V = \sum_{k=1}^{row}\sum_{l=1}^{col}\left(C_{kl,kl-1} + C_{kl,kl+1} + C_{kl,k-1l} + C_{kl,k+1l}\right) \to \max$$

with

$$C_{kl,rs} = \sum_i x_{ikl} \sum_j G_{ij} x_{jrs} \quad .$$

The basic structure of the graph G^* for this problem is shown in Figure 5.2.1.

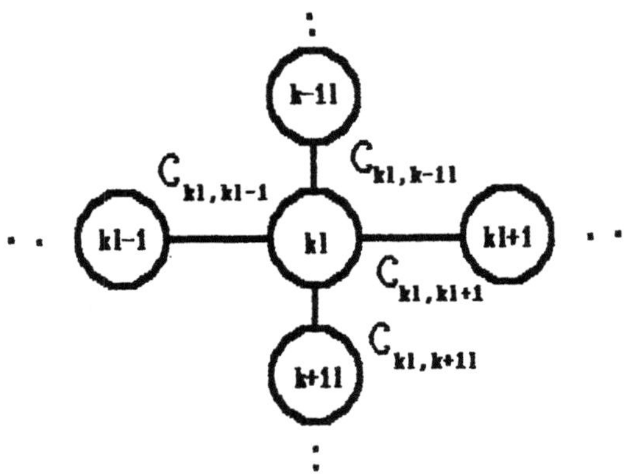

Figure 5.2.1 Basic structure of the graph G^* for the module placement problem

The Euclidean gradients are given by

$$\frac{\partial V}{\partial x_{ikl}} = \left(Gx_{.kl-1}\right)_i + \left(Gx_{.kl+1}\right)_i + \left(Gx_{.k-1l}\right)_i + \left(Gx_{.k+1l}\right)_i \quad .$$

Furthermore we have to consider the constraints

$$X_{kl}=\sum_i x_{ikl}=1 \quad .$$

Using the diversified replicator equation and introducing a selection pressure control u we have as fitness functions $f_{ikl}(x)$

$$f_{ikl}(x)=\frac{\sum_j G_{ij}\left(x_{jkl-1}+x_{jkl+1}+x_{jk-1l}+x_{jk+1l}\right)+u x_{ikl}}{X_{kl}} \quad \text{for } u\geq 0$$

$$f_{ikl}(x)=\frac{\sum_j G_{ij}\left(x_{jkl-1}+x_{jkl+1}+x_{jk-1l}+x_{jk+1l}\right)-u\left(1-x_{ikl}\right)}{X_{kl}} \quad \text{for } u<0$$

x_{ikl} is the probability with which module i is at the kth row and the lth column of the grid.

5.2.2. Example

We want to show the application of replicator network algorithms for the module placement problem by means of the example **TEST#31**. The 10 modules (vertices of the graph G) should be arranged on a grid with 2 rows and 5 columns. The problem was computed within 45 ... 85 iterations for 30 runs. For every run a minimal length placement was obtained at u = -1.0. The assignment x_{ikl} of modules to grid positions is shown in Figure 5.2.2. Figure 5.2.3 represents the corresponding placement.

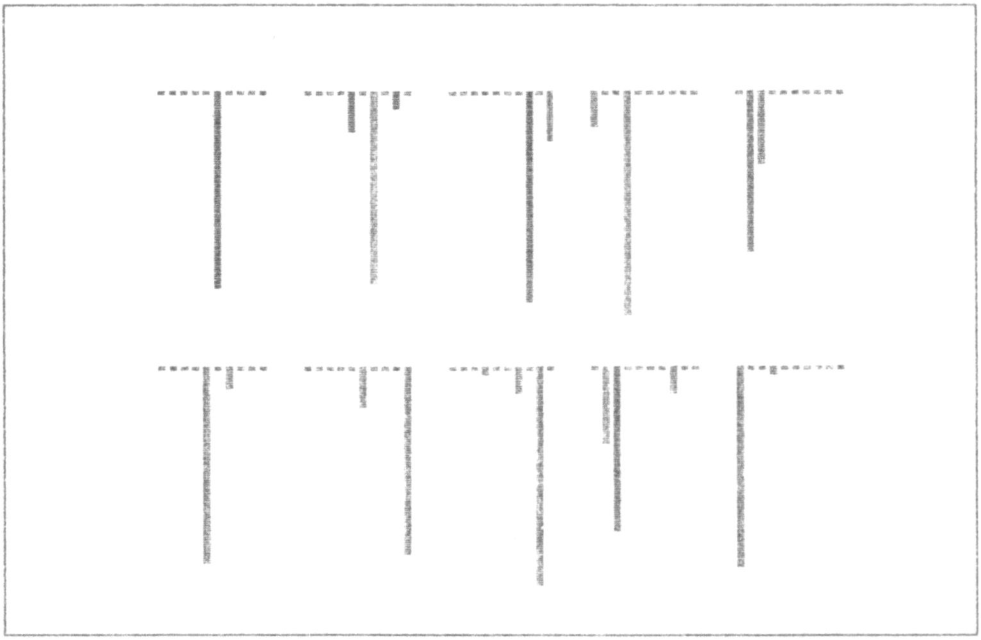

Figure 5.2.2 Assignment of modules to grid positions for example **TEST#31**

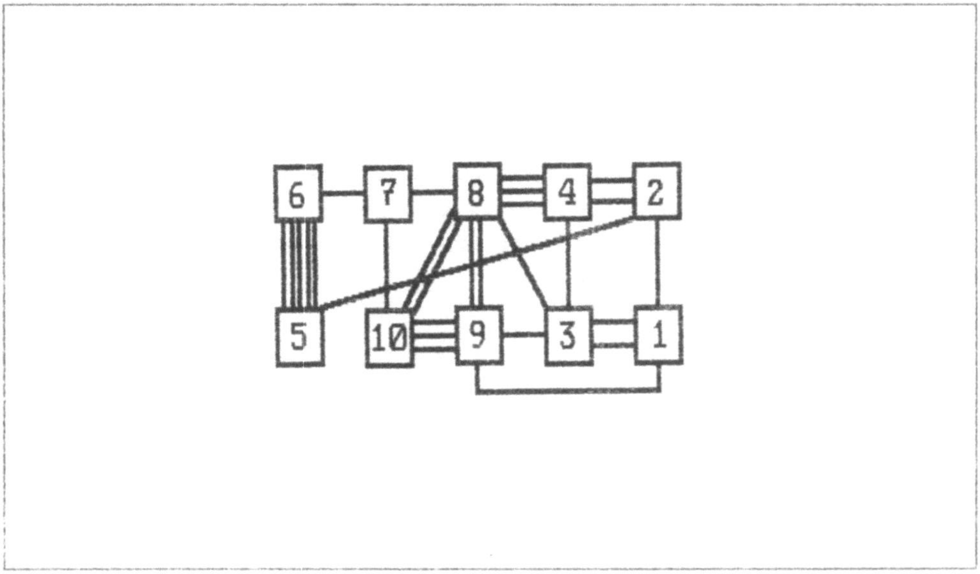

Figure 5.2.3 Placement of modules to grid positions for example **TEST#31**

5.3. The Travelling Salesman Problem

The formulation of the travelling salesman problem is quite simple and understandable.

Let us consider *n* cities. Every two cities are connected by a road with a certain length. Now, a travelling salesman wants to visit all cities once, and wants to return to the first city. The distance he travels should be minimal !

This problem is both challenging and interesting because it is so simple to state and so hard to solve. It is a so-called *NP*-hard combinatorial optimization problem.

NP-hard means, without digging too much into details, that no algorithm can be constructed which produces an exact solution and which can be done with an effort dependening only polynomially on the problem size. But this means that for the computation of exact solutions all possible combinations has to be enumerated.

How hopeless this approach will be shows the following example:

We consider 50 cities. The number of possible tours is

$$\frac{(n-1)!}{2}\text{ , i.e. } 3\cdot 10^{62} \text{ tours !}$$

Let us assume that the very powerful supercomputer SX-2 (NEC) with an operation speed of 1.2 Gigaflops (billions of floating point operations per second) is available. Furthermore we assume that for the computation of one tour only one floating point operation is necessary. Then we need for the computation of all tours for 50 cities with this supercomputer

$$7.5\cdot 10^{45} \text{ years .}$$

For comparison: Mankind exists some million years.

But to find good solutions for combinatorial optimization problems with manageable effort is important in a large number of application areas as e.g. integrated circuit design and artificial intelligence applications.

5.3.1. Fitness

The fitness functions for the above formulated travelling salesman problem can be formalized without any difficulties (Voigt & Santibanez-Koref (1988)). The edge weights $G_{ij} = G_{ji}$ of the graph G correspond to the distance between cities i and j for a suitable chosen distance measure. The vertices of G correspond to the cities.

Then the edge weights $C_{k,k+1}$ between adjacent super vertices of the graph G^* should be as small as possible. This yields the basic structure of the graph G^* for the travelling salesman problem depicted in Figure 5.3.1 with the potential function

$$V = \sum_k C_{kk+1} \to \min \ , \ C_{nn+1} = C_{n1}$$

with

$$C_{kl} = \sum_i x_{ik} \sum_j G_{ij} x_{jl} \ .$$

The Euclidean gradients are given by

$$\frac{\partial V}{\partial x_{ik}} = (Gx_{.k-1})_i + (Gx_{.k})_i + (Gx_{.k+1})_i \ .$$

Furthermore the constraints

$$X_k = \sum_i x_{ik} = 1$$

has to be taken into account. This will be done by penalty functions.

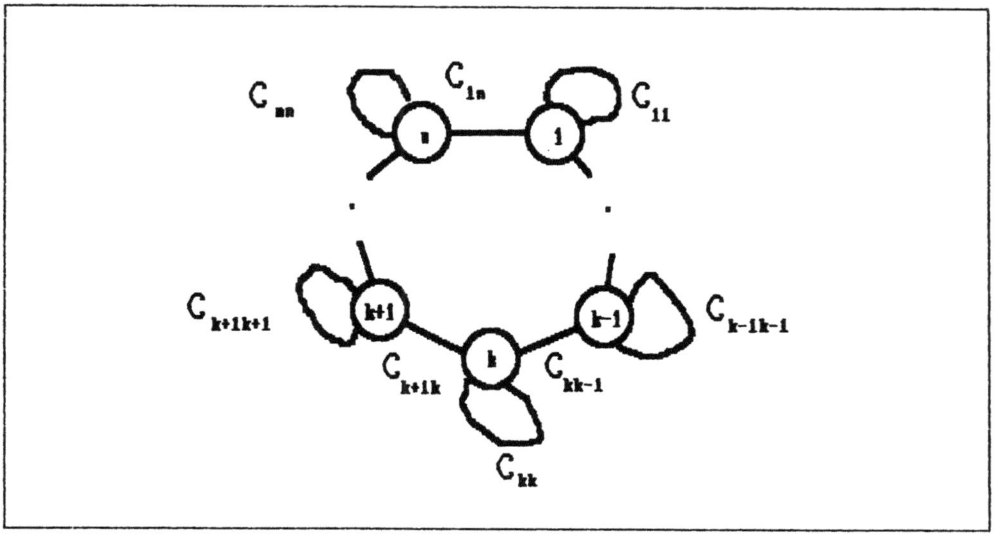

Figure 5.3.1 Basic structure of the graph G^* for the travelling salesman problem

Using an inverse diversified replicator equation corresponding to Section 2.12. because of the minimization goal and introducing the selection pressure control u we have as fitness functions $f_{ik}(x)$

$$f_{ik}(x) = \frac{\dfrac{1}{\sum_j c_{ij}(x_{jk-1}+x_{jk}+x_{jk+1})} + u x_{ik}}{X_k} \quad \text{for } u \geq 0$$

$$f_{ik}(x) = \frac{\dfrac{1}{\sum_j c_{ij}(x_{jk-1}+x_{jk}+x_{jk+1})} - u(1 - x_{ik})}{X_k} \quad \text{for } u < 0$$

x_{ik} is the probability with which city i is at the kth position of the salesman tour. The different treatment of positive and negative selection pressure can be done because of the invariance property for the continuous time replicator equation. It it necessary to ensure the simplex condition in case of the recurrence equation.

5.3.2. Example

We want to show the application of the replicator network algorithm for a small example which is taken from Hopfield & Tank (1985). The positions of cities are given by Table 5.3.1.

city #	x-position	y-position
1	.4000	.4439
2	.2439	.1463
3	.1707	.2293
4	.2293	.7610
5	.5171	.9414
6	.8732	.6536
7	.6878	.5219
8	.8488	.3609
9	.6683	.2536
10	.6195	.2634

Table 5.3.1 City positions of the Hopfield & Tank 10 city problem

We used a fixed selection pressure of u = -0.1 and a mutation rate h = 0.0001. For 20 runs we always found the optimal solution within 50 ... 80 iterations. The assignments for a typical run are shown for 40, 50, 60, and 70 iterations in Figure 5.3.2. The final tour including the tour length is presented in Figure 5.3.3.

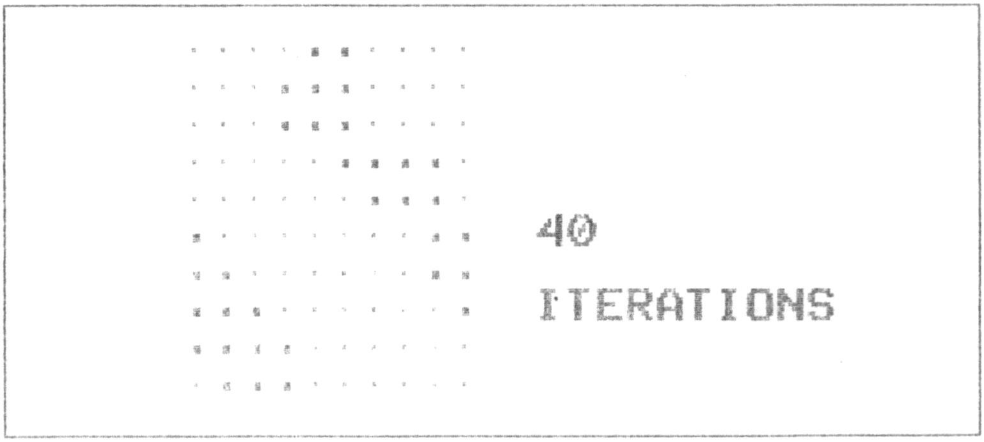

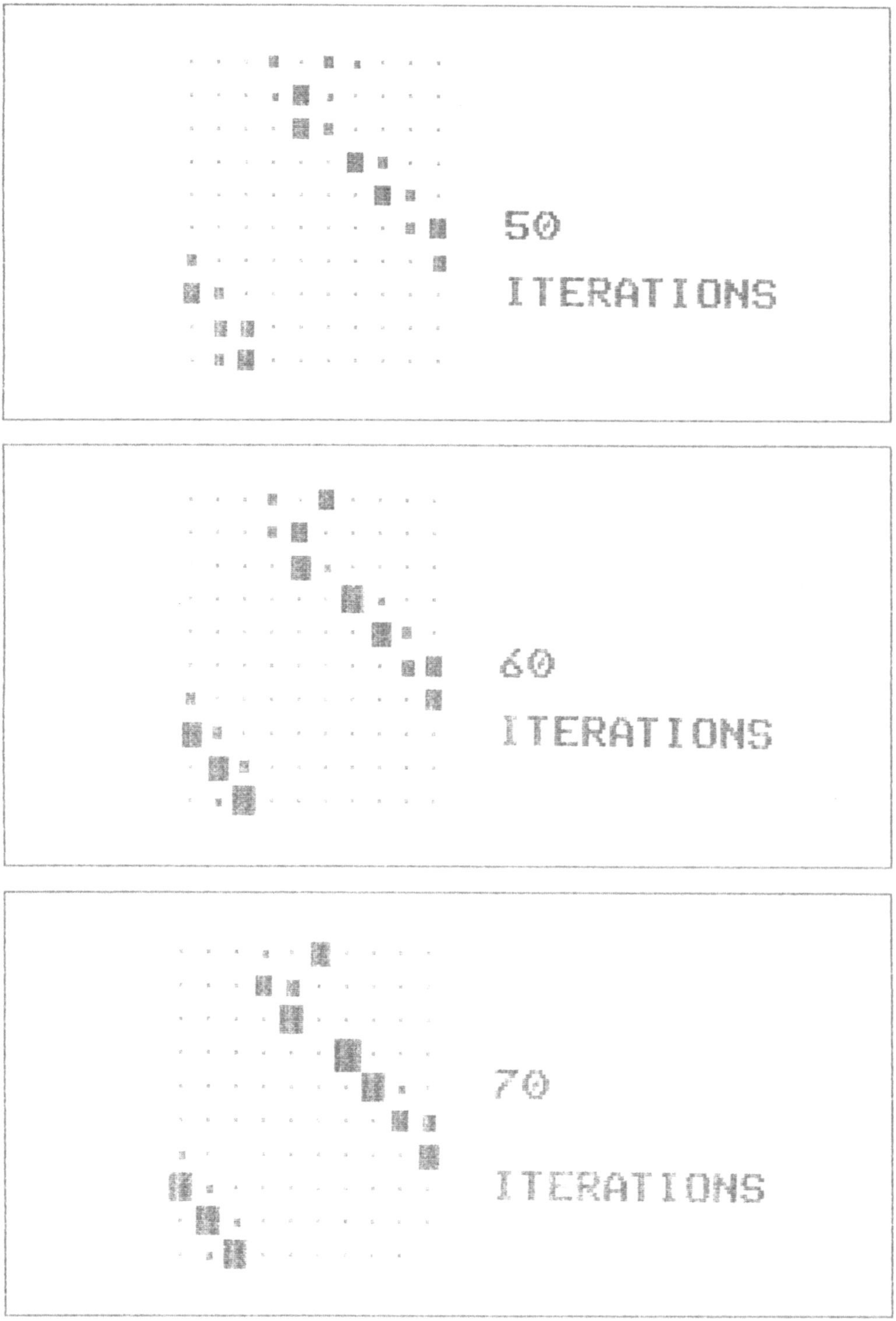

Figure 5.3.2 Assignments of cities (rows) to tour positions (columns) for the Hopfield & Tank 10 city example

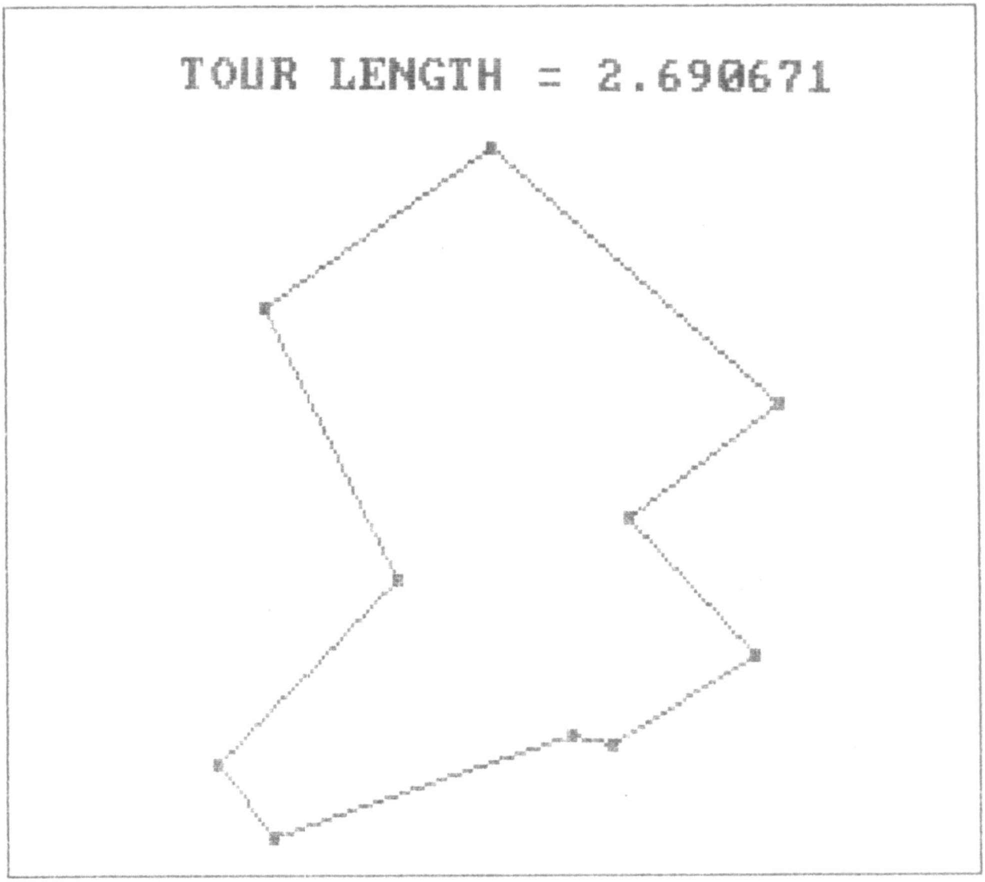

Figure 5.3.3 Final tour for the Hopfield & Tank 10 city example

A 20 city example is presented in Figures 5.3.4 and 5.3.5. The shown tour was found with 318 iterations, and is the minimal one.

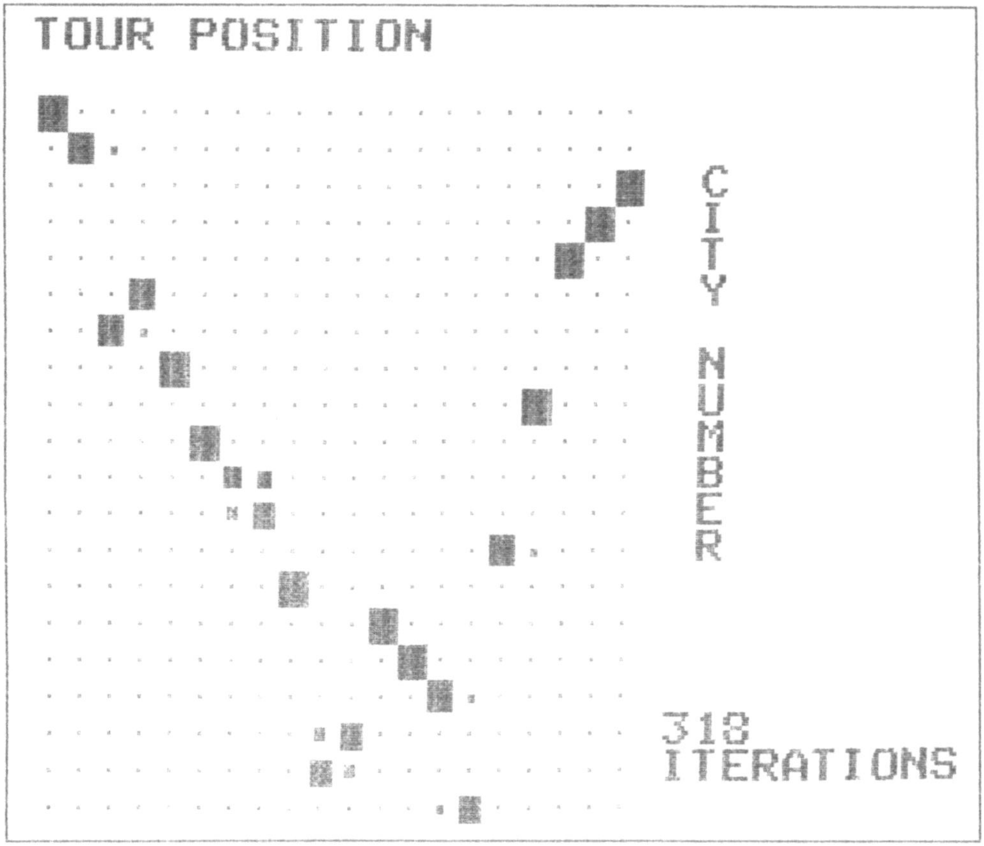

Figure 5.3.4 Assignments for the 20 city example

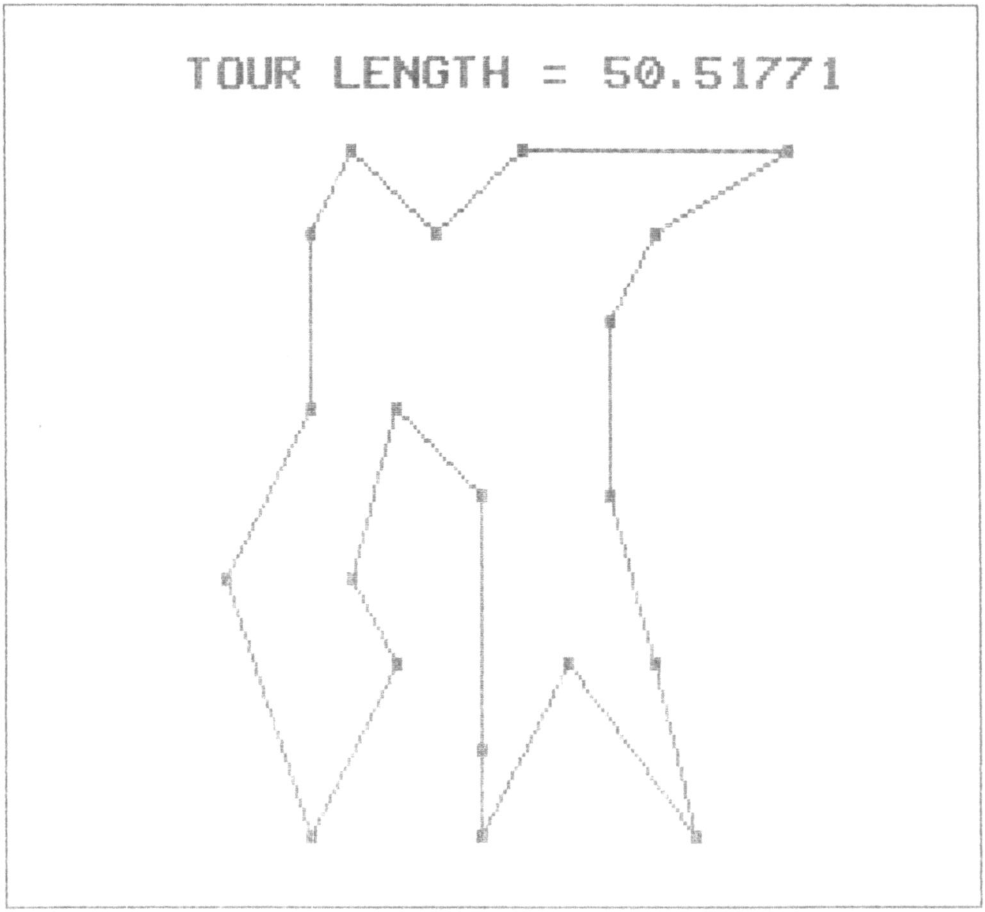

Figure 5.3.5 Tour for the 20 city example

We want to state some first conclusions comparing the replicator network approach with the Hopfield & Tank (1985) neural networks and with the analysis of the Boltzmann machine approach of Aarts & Korst.

Neural Networks (Hopfield & Tank)

Wilson & Pawley used the original function and parameter set of Hopfield & Tank to compute the Hopfield & Tank 10 city example. The simulation failed with the following results for 100 runs:

15 tours converged within 1000 iterations to valid tours (no one was optimal).
45 tours were frozen in configurations not corresponding to valid tours.
40 runs did not converge within 1000 iterations.

As we alreaay mentioned the optimal solution was always found by the replicator network algorithm.

Boltzmann machines (Aarts & Korts)

Aarts & Korts used for the solution of the Hopfield & Tank 10 city problem 5 min and for the 30 city problem 60 min CPU-time of a VAX-11/780.

With the replicator network approach the 10 city problem took 5 min and the 30 city problem 180 min computing time of a PC/XT-Compatible (8088 without coprocessor). Compared with the VAX 11/780 this is only a fraction of computing time.

6. Design of Integrated Circuits

Very large scale integrated circuits (VLSI) represent the most advanced form of electronic systems.

We speak of systems, because central processing units (CPU) with wordsizes of 32 bits and more can be integrated nowadays on one or on only a few chips. Random access memories (RAM) with a storage density of one megabit are already shipped, memory chips with much higher storage density are under development.

The design of such systems is characterized by a rapid increase of complexity and difficulty with increasing integration level, and technological level, respectively.

The difficulty of the design and the production of such circuits is emphasized by the fact that almost all disciplines from natural sciences, technical sciences, and mathematics have to work together. This concerns especially electrical and electronics engineering, computer science, cybernetics, mathematics, physics, and chemistry.

This variety will yet be multiplied if one takes into account the increasing migration of microelectronics, especially of application specific integrated circuits (ASIC), into the different application areas.

Quite typical is the increasing complexity with increasing integration level. This fact becomes clear from the following example (Losleben (1982)).

In the early seventies metal connection widths of 10 micron and a distance between connections of 7.5 micron were quite usual. Hence, a wiring grid for the connections of 17.5 micron results. Let us assume a design grid of 6.354 mm (1/4 inch) for representing the wiring grid. A chip of 6.35 mm x 6.35 mm (250 mil) area can then be drawn on a sheet of 2.26 m x 2.26 m = 5.1 m^2 (7.4 ft square) at a scale of 357x. decreasing the metal connection width for the same chip area and using the same design grid yields the complexity evolution depicted in Table 6.1.1.

Year	Grid Size micron	Scale	Paper Size m^2
early 70	17.5	357	5.1
early 80	8.0	800	26.8
mid 80	2.0	3200	417.0
future	0.4	16000	1030.0

Table 6.1.1 Evolution of circuit design complexity

For metal interconnect width of 0.4 micron the drawing paper has the size of two football fields. You must imagine now a team of designers crawling on two football fields and filling it with a circuit layout using a design grid of 6.35 mm !

Finally it should be noted that within this section we want to show problems and possible solutions in the context of replicator equations. It is not intended to present an exhaustive course on circuit design.

6.1. Design Goals

Corresponding to the projected evolution especially of circuit design complexity a number of design goals can be deduced. This concerns

(i) Design cycle time

(ii) Design effort

(iii) Design performance

These global goals can be detailed e.g. in the following form

(i) Number of designed elements per designer and day

(ii) Number of design errors for a circuit

(iii) Technical and technological parameters

- Used chip area
- Power dissipation

- Switching speed of elements
- Clock frequency of circuit
- Testability

These goals can only be met by an overall computer support. This reason for, extraordinary efforts are made to develop design systems for integrated circuits. Such systems can be considered as prototype systems for computer aided design applications because of the complexity and difficulty of very large scale integrated circuit design. A design system has to be considered as a unity of

(i) Hardware

(ii) Software

(iii) Technology (design, production),

i.e. as a dedicated work station.

Features of these components can best be characterized by the following attributes

(i) Computer hardware

- High working speed
- Large direct access memory
- Direct access to fast peripheral mass memory
- Direct connection of high performance graphical peripherals

(ii) System and application software

- Operating system with effective man machine interaction support, and effective file and library functions
- Data management system
- Interaction frame systems
- Text and graphics editors
- Design simulators
- Test pattern generators
- Layout generators
- etc.

(iii) Design technology

- Standard circuits
  Processors

Memories
etc.
- Application specific circuits
Gate arrays
Standard cells
Macro cells
etc.

This necessarily incomplete list shows the diversity of the development of design systems for integrated circuit design.

6.2. Design Levels and Design Tools

Because of the complexity in circuit design the above mentioned goals can be met only if the designer can work with the necessary (not with the possible) details. This implies that a hierarchical description of the circuit design must be supported.

At every description level the corresponding tools must be at hand to complete, to correct, or to optimize a design by analysis or synthesis. A problem with this design style is to ensure the consistency of the design data for different design levels and different design phases. One way to overcome this problem consists in the utilization of suitable integrated data management mechanisms. Furthermore, dependent on the design technology, the transition between the design levels in a top down as well as a bottom up way must be possible. By this, the simulation (top down) and the verification (bottom up) will be supported. The test pattern generation is necessary for all design levels.

These relations are shown in Figure 6.2.1.

Editors for	Text and graphics	Database	Data management	Test patterns	Design rules	Simulation	Levels	Verification
							System level RTL description System simulators	
							Logic level Logic description Logic simulators (static/dynamic)	
							Electric level Network description Network analysis	
							Layout level Layout description (symbolic/physical) Layout generators	

Figure 6.2.1 Hierarchies for the design of complex circuits

As an example we want to look at the logic level in somewhat more detail. For the dynamic two and three value logic simulation the simulation package **DICSI** (DIgital Circuit SImulation and Analysis Program) was developed at the Institut für Regelungstechnik (Voigt (1977)). This system was applied successfully for the simulation of LSI circuits, so e.g. for the timer IC's **D355** and **D350**. The circuits were realized in I^2L technology and are in use especially for electronic time relays with a broad application range.

The description of circuits with the **DICSI** system is done by means of the language **DIC** (Description of Integrated Circuits) (Voigt (1978)).

Problems and possibilities of the generation and optimization of test patterns at the logic level are presented in Voigt (1980). The optimization of the test patterns was also used for the circuit **D355**.

It should be noted that during the last ten years decisive joint efforts were made in the GDR by the microelectronics industry, the Akademie der Wissenschaften der DDR, and universities to develop new, more powerful software systems for the design of integrated circuits.

A typical system for the design of application specific integrated circuits is the CMOS gate array system U 5200 of the Centre for Microelectronics at Dresden. This system can be characterized by the following parameters:

CMOS gate array system U 5200

- cSGT technology
- 3 programming levels
- 3000 logic gates (corresponds to 12000 transistors)
  with 1000 logic gates
  100 JK master slave flip flops
  (1 flip flop corresponds to 20 gates)
- Typical delay time t_D = 5 ns
  (standard NAND 2)
- Clock frequency t_C = 10 MHz
- 64 pins (CC 64)
- Approximately 100 macros
- Design software
  Compiler for NBS 84
  Dynamical logic analysis
  Placement
  Global routing
  Channel routing
  Test pattern generation (LSSD)
  Pattern conversion
  Data management
  Test management

6.3. Design Goals and Fuzziness

At the beginning of an idealized top down design one realizes that only a more or less verbal description of the circuit is available. This description is usually rather unformalized. It belongs almost exclusively to a semantic level. The description contains the design goals in a highly aggregated form as e.g. the realization technology is because of several reasons CMOS ASIC. In the course of the definition and design process the units of the semantic level will be more and more formalized and supported by syntactic description tools. So e.g. CMOS ASIC can be replaced by a gate array circuit based on the U 5200 system, which implies a number of further decisions, or by full custom ASIC with cSGT2 technology.

But this means during a definition and design process we have to follow a decision tree. The aim for the decision process is to run through the tree or parts of it only once to meet the goals of design cycle time and design effort.

This reason for, in the course of a design process decisions should be made in such a way that the range of subsequent decisions will be constraint only as far as necessary.

So the fuzziness inherent to decision processes should not only be obeyed but should be used goal oriented.

From a methodological point of view fuzziness can be included via probabilistic models or via fuzzy sets theory.

We want to clarify the goal oriented inclusion of fuzziness by means of layout design problems.

The layout design consists roughly spoken in a transformation of a design on the logic or electrical network level, which was functionally and structurally designed and checked at this level, to the constructive form of the layout geometry. The constructive parameters of the layout are given by the technology dependent design rules.

To overcome the complexity barrier different descriptive and algorithmic levels has to used. This concerns especially

(i) Partitioning

(ii) Global placement, and floorplanning

(iii) Final placement

(iv) Global wiring

(v) Final wiring

Dependent on the problem different methods and algorithms can be used. So e.g. for final wiring purposes channel routers, swith box routers, or Lee routers.

But it has to be clear that fixations at one level have decisive influence on the solution performance of the next lower levels. At the same time it has to be clear that all goals cannot be aggregated in only one performance measure.

We want to show by an example how fuzziness in graph partitioning can be used for global placement. For this purpose we consider the network depicted in Figure 6.3.1 (Götze & Nehrlich (1985)) with the shown arbitrary placement.

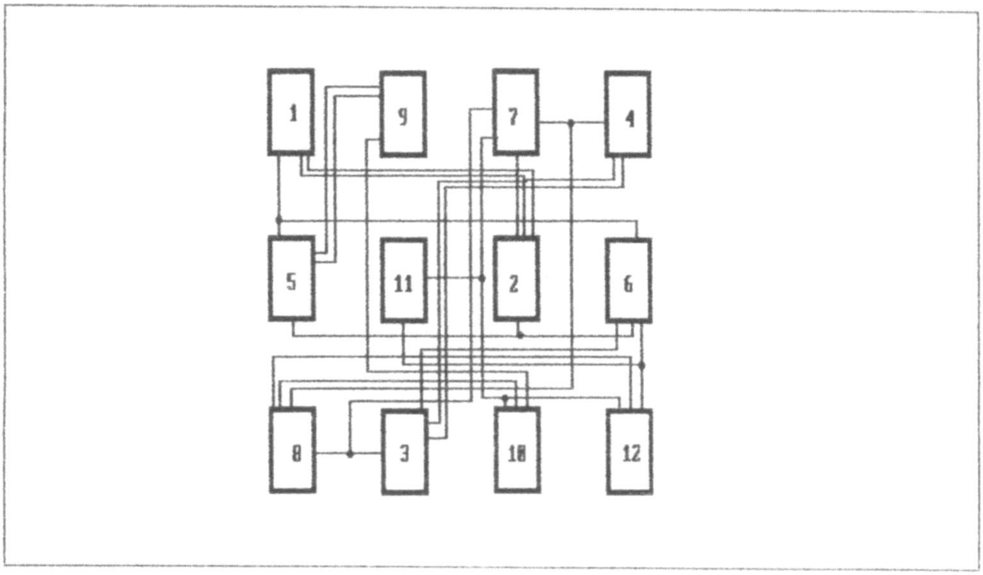

Figure 6.3.1 Example

Now we decompose it into m = 3 partitions with a balance control a = 0 and a selection pressure schedule u = -1.0, -0.8, -0.5. Then we have the module assignments to different partitions as shown in Figure 6.3.2.

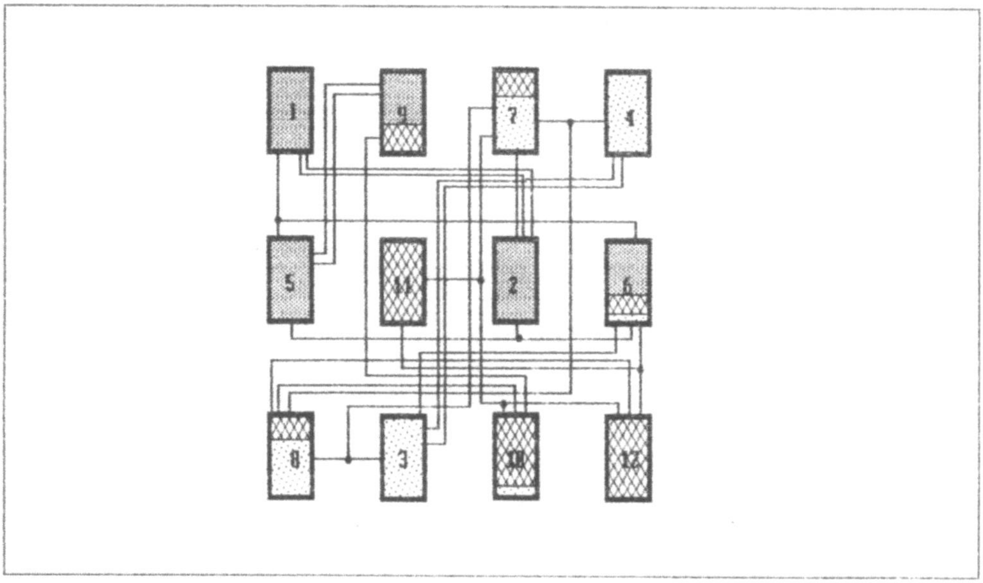

Figure 6.3.2 Assignment of modules to partitions

From this partitioning we can at once deduce possible placement methods:

(i) Interactive placement

We assume that for this task a workstation with interactive high-resolution colour graphics is used. Then we can label different partitions by different colours. The membership of modules to different partitions can be shown by the corresponding percentage area and colour within each module. So the designer is shown the adjacency of modules because of their connection structure. By an interactive interchange of modules on the screen the desired placement can be generated. Possible constraints can be checked and signalled easily for the interchange operations. For the example a placement is shown in Figure 6.3.3 for fixed positions of the modules 1, 4, and 12.

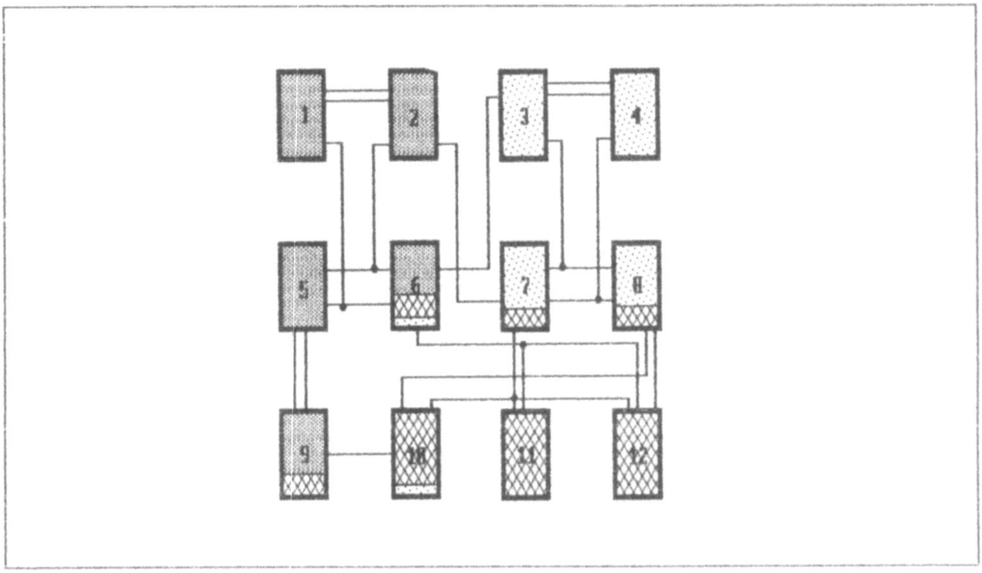

Figure 6.3.3 Placement of modules by interactive interchange

(ii) Interactive floorplanning with expert system support

The interchange operations including constraints can be of course supported by an expert system. But such a support can be extended. So e.g. if the detection of characteristic structures will be supported for which dedicated wiring algorithms can be applied. Such characteristic structures are e.g. data paths, arrays, channels, circuits, simple rows, and so-called H-clusters corresponding to Figure 6.3.4 (Daes (1985)). This direction of design support is especially interesting in connection with silicon compilation.

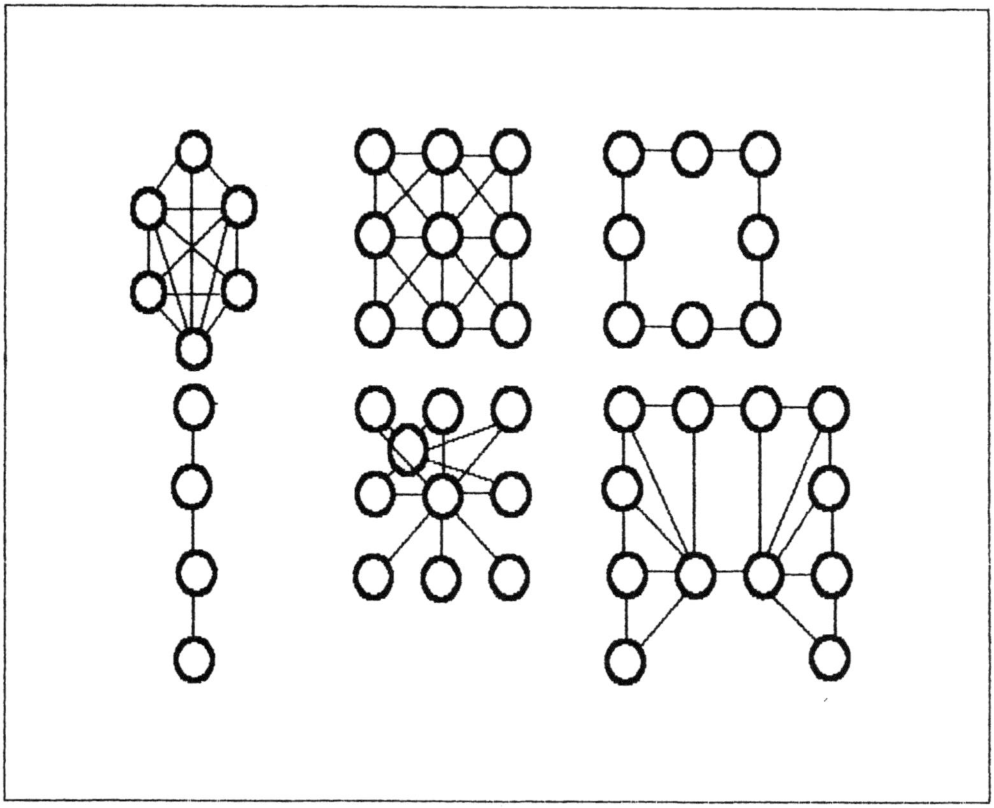

Figure 6.3.4 Different characteristic structures

7. Mapping Strategies for Multiprocessor Systems

Connected with the enormous advancement of microelectronics one recognizes a rapid progress in computer architecture development.

The main characteristics of new computer architectures are an increasing parallelism and fault-tolerance coupled with diversified functionality. As Hillis (1985) pointed out "The problem lies in how to couple the raw power with the applications of interest, how to program the hardware to the job.".

That concurrency offers a solution can be best seen by the performance the human brain works though single neurons are very slow processing units. Obviously, the arrangement of the processing elements corresponds to the natural structure of the data.

The difference of approaches and underlying principles of computer architectures should be shown by some examples of innovative computer structures and functions:

(i) A very fine grained, highly parallel system is the Connection Machine of Thinking Machines (Hillis (1985)) which is especially dedicated to image processing, VLSI simulation, and semantic networks. The CM-1 prototype Connection Machine consists of 65536 (64 K) very simple processing elements each with 4096 bits of memory and realizes a parallelism on the data level.

(ii) The ANZA coprocessor board for IBM PC's of Hecht-Nielsen Neurocomputer Co. makes it possible to implement different neural network concepts containing up to 30000 neurons with up to 480000 interconnects. It is especially designed for speech and image recognition.

(iii) Transputers were developed by INMOS for parallel computer applications. A transputer chip contains a 32 bit arithmetic logic unit (ALU), a floating-point coprocessor, 4096 bytes of local memory, and 4 I/O ports. By means of the OCCAM programming environment transputers can be arranged to meet the specific applications. This architecture is especially designed to realize Hoare's (1978) concept of communicating sequential processes.

(iv) The SUPRENUM supercomputer (Behr, Giloi & Mühlenbein (1986)) will be designed especially for numerical simulation with multigrid methods. It consists of clusters connected by horizontal and vertical ring busses. Each cluster contains a number of node computers containing a 32 bit microprocessor, a 64 bit floating point coprocessor, 2 megabyte local memory, and a communication unit.

In the next sections we want to show the application of the replicator network approach to the mapping of processes to processors for message-based multiprocessor systems. The application of the graph partitioning algorithm (Voigt (1984,1986)) to the mapping problem was first done by Mühlenbein, Gorges-Schleuter & Krämer (1987).

7.1. Local Memory Abstract Machines

In analyzing and creating new computer architectures one has to regard a number of tradeoffs. So, different viewpoints as e.g. the operation principle are emphasized. Here we want to distinguish between multiprocessor systems with

(i) shared memory

(ii) distributed memory.

Both system types have a problem to be solved for efficient application.

With shared memory multiprocessors the synchronization of memory access represents the essential bottleneck. Such systems consists of a number of processors all working on common data in a common memory. It is quite clear that processors working on the same data at the same time will generally conflict in data access.

This problem will be avoided by distributed memory multiprocessors where each processor has its own local memory. But obviously, data produced by a process and stored in a corresponding local memory may be needed by other processes. The distribution of data via a suitable communication system represents the main problem of such multiprocessor systems.

This has to be considered as an assignment of processes to processors such that the communication costs are minimized. It depends on the concrete communication structure. The most prominent communication structures with this respect are

(i) Rings
(ii) Lattices
(iii) Trees
(iv) Hypercubes.

In the next section we consider mapping strategies for distributed memory multiprocessor systems.

7.2. Mapping of Process Nets to Processor Networks

In developing mapping strategies for multiprocessor systems, i.e. for the assignment of processes to processors, we consider the following issues

(i) static process structure

(ii) fixed communication structure.

With these assumptions, a distributed memory architecture is defined in a simplified way by the following two graphs:

(i) Process graph $G = (V,E)$

V denotes the set of processes and E the set of existing communications between processes. The edge values G_{ij} represent the communication amount from process j to process i.

(ii) Processor graph $G^* = (V^*,E^*)$

V^* denotes the set of processors and E^* the set of communication links. The edge values G^*_{kl} represent the communication costs per communication unit between processor k and processor l.

We assume now the assignment probabilities x_{ik} of process i to processor k, and x_{jl} of process j to processor l. Then the communication costs between process i on processor k and process j on process l are given by

$$C_{ik,jl} = G^*_{ij} x_{ik} G_{ij} x_{jl} \quad .$$

The total communication costs between processor k and processor l are given by

$$C_{kl} = \sum_i x_{ik} \sum_j G^*_{ij} G_{ij} x_{jl} \quad .$$

The goal consists now in determining such assignments of processes to processors that the total communication costs are minimal. The application of such an approach depends strongly on the processor network structure G^*. Hence we show the application for two such network structures.

Structure 1: Clustered networks

Let us assume that processors are arranged in a cluster with a very fast bus corresponding to Figure 7.2.1. The clusters are connected by horizontal and vertical ring busses to a torus as it is shown in Figure 7.2.2. This is the principal structure of the SUPRENUM supercomputer.

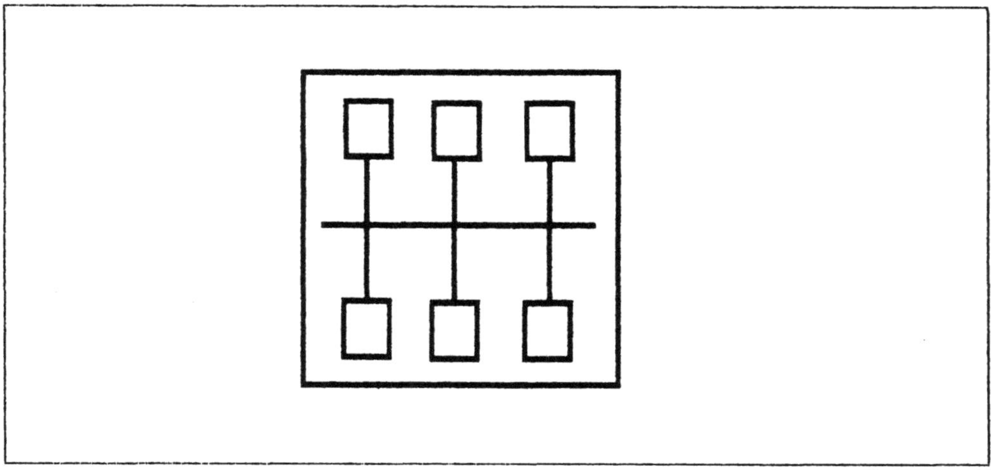

Figure 7.2.1 Cluster of processors

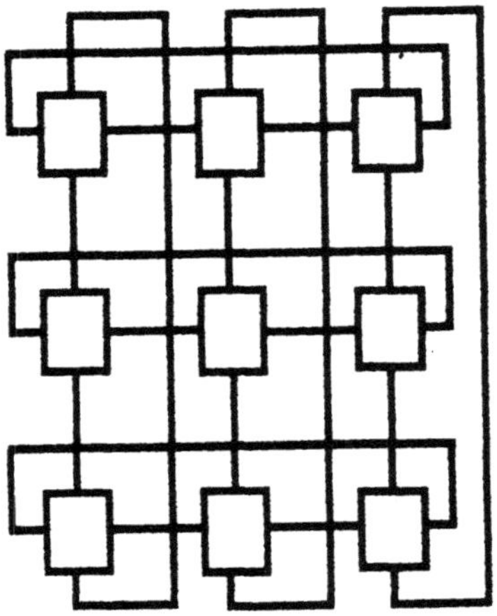

Figure 7.2.2 Network of processor clusters

The communication costs per communication unit within one cluster should be small compared with the communication costs between clusters. Then it is quite natural to arrange processes with a high mutual communication amount within one cluster. The assignment of the process clusters to the concrete processor clusters can then be done randomly. Hence, the graph partitioning algorithm from Section 5.1. can be applied to the process graph G with respect to the number of given processor clusters. Since the amount of the communication costs plays the dominating role we use as the appropriate potential function

$$V = \prod_k C_{kk} \rightarrow \max$$

The efficiency of such a partitioning algorithm was shown in Section 5.1.

Mühlenbein, Gorges-Schleuter & Krämer (1987) showed the near optimality of the overall mapping by such a partitioning.

Structure 2: Torus networks

Let us assume now a torus network with single processing elements having 4 I/O-links to the adjacent neighbours. Such elements are e.g. given by the INMOS transputer chips. The arrangments are shown in Figures 7.2.3 and 7.2.4.

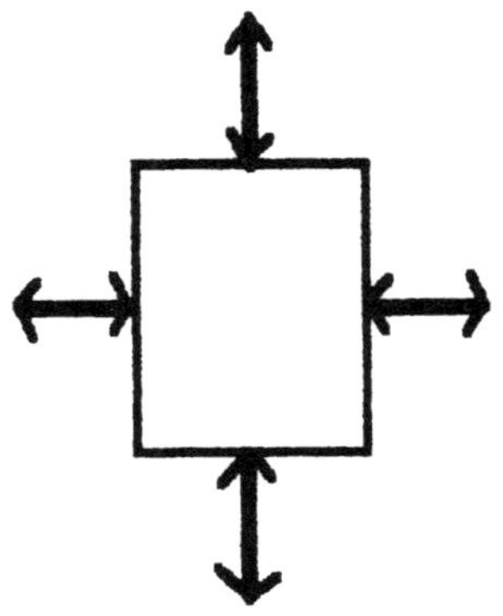

Figure 7.2.3 Processor with 4 I/O-links

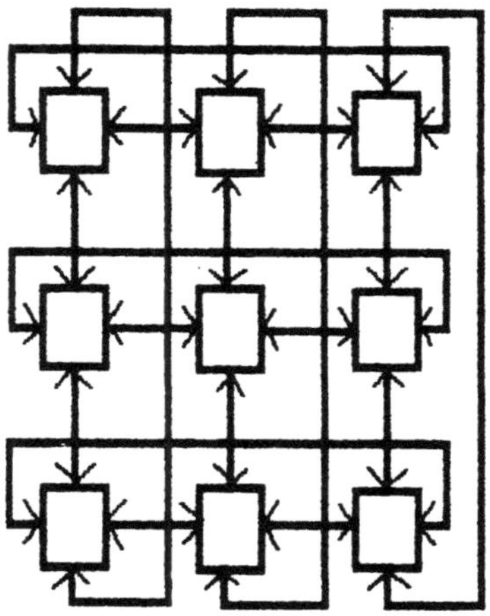

Figure 7.2.4 Torus network of processors

Obviously, the communication costs between adjacent processors are less than the costs between non-adjacent processors. Hence, processors with a large amount of mutual communication should be adjacent. This corresponds to the module placement problem from Section 5.2. with the specific torus structure. In this case the potential function is given by

$$V = \sum_{k=1}^{row}\sum_{l=1}^{col}\left(C_{kl,kl-1} + C_{kl,kl+1} + C_{kl,k-1l} + C_{kl,k+1l}\right) \to \max$$

with k = k mod(row) and l= l mod(col).

For other network structures, such as e.g. hypercubes, suitable mapping strategies can be derived from replicator network algorithms in much the same way.

8. References

Aarts, E.H.L. & P.J.M. van Laarhoven (1985): Statistical Cooling: A General Approach to Combinatorial Optimization Problems. Philips J. Res. **40**, 193-226

Aarts, E.H.L. & J.H.M. Korst (1987): Solving Travelling Salesman Problems with Massively Parallel Networks Based on the Boltzmann Machine. European Journal of Operations Research (to appear)

Abley, P. (1987): Optimieren mit Evolutionsstrategien. Spektrum der Wissenschaft, Juli 1987, 104-115

Akin, E. (1979): The Geometry of Population Genetics. Lecture Notes in Biomathematics **31**. Berlin, Heidelberg, New York: Springer Verlag

Amann, E. & J. Hofbauer (1985): Permanence in Lotka-Volterra and Replicator Equations. Proc. 5th Meeting UNESCO's Working Group on System Theory 1984, 23-34, Berlin: Akademie-Verlag

Amari, S. & M.A. Arbib (Ed.) (1982): Competition and Cooperation in Neural Nets. Lect. Notes in Biomathematics **45**. Berlin, Heidelberg, New York: Springer Verlag

Arneodo, A., P. Coullet & C. Tresser (1980): Occurrence of Strange Attractors in Three Dimensional Volterra. Equations. Physics Letters **79A**, 259-263

Ayres, R.F. (1983): VLSI: Silicon Compilation and the Art of Microchip Design. Englewood Cliffs: Prentice Hall

Behr, P.M., W.K. Giloi & H. Mühlenbein (1986): SUPRENUM: The German Supercomputer Project - Rationale and Concepts. IEEE Int. Conf. on Parallel Processing

Bomze, I.M. (1986): Non-Cooperative Two-Person Games in Biology: A Classification. Int. J. Game Theory, vol. **15**, no. 1, 31-57

Born, J. (1985): Adaptively Controlled Random Search - A Variance Function Approach. Systems Analysis, Modelling and Simulation **2**, 109-120

Boseniuk, T., W. Ebeling & A. Engel (1988): Boltzmann und Darwinian Strategies in Complex Optimization. Physics Letters (to appear)

Brady, R.M. (1985): Optimization Strategies Gleaned from Biological Evolution. Nature vol. **317**, 804-806

Braun, M. (1979): Differentialgleichungen und ihre Anwendungen. Berlin, Heidelberg, New York: Springer-Verlag

Breuer, M.A. (Ed.) (1972): Design Automation of Digital Systems: Theory and Techniques. Englewood Cliffs: Prentice-Hall

Breuer, M.A. (Ed.) (1975): Digital System Design Automation: Languages, Simulation, and Data Base. Potomac: Computer Science Press

Casti, J.L. & A. Karlqvist (Ed.) (1986): Complexity, Language and Life: Mathematical Approaches. Berlin, Heidelberg, New York: Springer Verlag

Cerny, V. (1983): A Thermodynamic Approach to the Travelling Salesman Problem: An Efficient Simulation Algorithm. J. Opt. Th. and Appl.

Cohoon, J.P. & W.D. Paris (1987): Genetic Placement. IEEE Trans. on Computer Aided Design of Integrated Circuits and Systems, **CAD-6**, no. 6, 956-964

Crow, J.F. & M. Kimura (1970): An Introduction to Population Genetics. New York: Harper and Row

Daes, A.R. (1983): The Unit Silicon Compiler. University of Edinburgh, CSR-145-83

Daes, A.R. (1985): An Expert System for VLSI Floorplanning. University of Edinburgh

Darwin, Ch. (1859): On the Origin of Species by Means of Natural Selection. Cambridge, London (1964): Harvard University Press (Reprint), Stuttgart (1906): Alfred Knöner Verlag (German)

Dawkins, R. (1976): The Selfish Gene. Oxford: Oxford University Press

Dawkins, R. (1982): The Extended Phenotype. Oxford, San Francisco: W.H. Freeman

Denker, J. (Ed.) (1986): Computing with Neural Networks. American Institute of Physics

Deo, N. (1974): Graph Theory with Applications to Engineering and Computer Science. Englewood Cliffs: Prentice-Hall

Durbin, R. & D. Willshaw (1987): An Analogue Approach to the Travelling Salesman Problem Using an Elastic Net Method. Nature **326**, 689-691

Ebeling, W. & R. Feistel (1982, 1986): Physik der Selbstorganisation und Evolution. Berlin: Akademie-Verlag

Ebeling, W. & A. Engel (1986): Models of Evolutionary Systems and Their Application to Optimization Problems. Syst. Anal. Model. Simul. vol. 3, no. 5

Eigen, M. & P. Schuster (1979): The Hypercycle: A Principle of Natural Selforganization. Berlin, Heidelberg, New York: Springer-Verlag

Ewens, W.J. (1979): Mathematical Population Genetics. Berlin, Heidelberg, New York: Springer-Verlag

Faddejev, D.K. & W.N. Faddejeva (1964): Numerische Methoden der linearen Algebra. Berlin: VEB Deutscher Verlag der Wissenschaften

Fisher, R.A. (1930): The Genetical Theory of Natural Selection. Oxford: Clarendon Press

Freeman, E. (1985): Computer Aided Engineering/Design Workstations. EDN, May 30, 141-156

Garey, M.R. & D.S. Johnson (1979): Computers and Intractability: A Guide to the Theory of NP-Completeness. San Francisco: W.H. Freeman

Gössel, M. (1972): Angewandte Automatentheorie. WTB Mathematik-Physik. Berlin: Akademie-Verlag

Götze, B. & W. Nehrlich (1985): Some Placement Techniques. Berlin: Acad. Sci. GDR, Inst. for Informatics, IIR-Report vol. 1, no. 12, 97-114

Goldberg, D.E. & R. Lingle Jr. (1985): Alleles, Loci, and the Traveling Salesman Problem. Proc. Int. Conf. on Genetic Algorithms and their Applications, 154-159

Gorges-Schleuter, M. & H. Mühlenbein (1988): Die Evolutionsstrategie - Prinzip für parallele Algorithmen. Jahresbericht der Gesellschaft für Mathematik und Datenverarbeitung 1988, St. Augustin

Grefenstette, J., R. Gopal, B. Rosmaita & D. van Gucht (1985): Genetic Algorithms for the Traveling Salesman Problem. Proc. Int. Conf. on Genetic Algorithms and their Applications, 160-168

Grossberg, S. (1982): Associative and Competitive Principles of Learning and Development. Lect. Notes in Biomathematics **45**. Berlin, Heidelberg, New York: Springer Verlag, 295-341

Gumowski, I. & C. Mira (1980): Recurrences and Discrete Dynamic Systems. Berlin, Heidelberg, New York: Springer-Verlag

Hadeler, K.P. (1974): Mathematik für Biologen. Berlin, Heidelberg, New York: Springer-Verlag

Haken, H. (1978): Synergetics: An Introduction. Berlin, Heidelberg, New York: Springer-Verlag

Haldane, J.B.S. (1932): The Causes of Evolution. New York: Harper and Row

Hayes, B. (1984): Computer Recreations: The Cellular Automaton Offers a Model of the World and a World unto Itself. Scientific American 3, 10-16

Hillis, W.D. (1985): The Connection Machine. Cambridge: The MIT Press Series in Artificial Intelligence

Hoare, D. (1978): Communicating Sequential Processes. Comm. of the ACM, vol. **21**, no. 8, 666-677

Hofbauer, J. (1981): On the Occurrence of Limit Cycles in the Lotka-Volterra Equation. Nonlinear Analysis vol. **5**, no. 9,1003-1007

Hofbauer, J. & K. Sigmund (1984): Evolutionstheorie und dynamische Systeme. Berlin, Hamburg: Paul Parey

Hofbauer, J. (1985): The Selection Mutation Equation. IIASA Collaborative Paper CP-85-2

Holland, J.H. (1975): Adaption in Natural and Artificial Systems. Ann Arbor: University of Michigan Press

Hopfield, J.J. & D.W. Tank (1985): Neural Computation of Decisions in Optimization Problems. Biol. Cybernetics **52**, 141-152

Hopfield, J.J. & D.W. Tank (1986): Computing with Neural Circuits: A Model. Science, vol. **233**, no. 4764, 625-633

Jansen, W. (1987): A Permanence Theorem for Replicator and Lotka-Volterra Systems. J. Math. Biol. **25**, 411-422

Kernighan, B.W. & S. Lin (1970): An Efficient Procedure for Partitioning Graphs. Bell System Techn. J., 291-307

Kingman, J. (1961): A Matrix Inequality. Quarterly J. Math. vol. **12**, 78-80

Kirkpatrick, S., C.D. Gelatt Jr. & M.P. Vecchi (1983): Optimization by Simulated Annealing. Science vol. **220**, no. 4598, 671-680

Kolmogoroff, A. (1930): Sur La notion de La moyenne. Atti della Reale Academ. nation. dei Lincei, Rend. adunance solenni, ser. 6, 12, 388-391

Krämer, O. & H. Mühlenbein (1987): Mapping Strategies in Message Based Multiprocessor Systems. Lect. Notes in Computer Science **258**. Berlin, Heidelberg, New York: Springer Verlag, 213-225

Kubicek, M. & M. Marek (1983): Computational Methods in Bifurcation Theory and Dissipative Structures. Berlin, Heidelberg, New York: Springer-Verlag

Kurka, P. (1985): Evolution of Replicators Playing a Strategic Game. Biological Cybernetics **52**, 211-217

Landau, L.D. & EM Lifschitz (1975): Lehrbuch der theoretischen Physik. Teil V Statistische Physik. Berlin: Akademie-Verlag

Lawlor, E.L. (1976): Combinatorial Optimization. New York: Holt, Reinhart and Winston

Lawlor, E.L., A. Lenstra & H.G. Rinnooy Kan (1985): The Travelling Salesman Problem. New York: John Wiley

Lin, S. (1965): Computer Solutions of the Travelling Salesman Problem. Bell System Techn. J. **44**, 2245-2269

Lippman, R.P. (1987): An Introduction to Computing with Neural Nets. IEEE ASSP Magazine, April 1987, 4-22

Losleben, P. (1982): Computer Aided Design for VLSI. In: D.F. Barbe (Ed.): Very Large Scale Integration: Fundamentals and Applications. Berlin, Heidelberg, New York: Springer-Verlag, 89-127

Lotka, A.J. (1920): Analytical Note on Certain Rhythmic Relations in Organic Systems. Proc. N.A.S. vol. **6**, 410-415

Marsden, J. & M. McCracken (1976): The Hopf Bifurcation and its Applications. Berlin, Heidelberg, New York: Springer-Verlag

Maynard Smith, J. & G.R. Price (1973): The Logic of Animal Conflicts. Nature vol. **246**, 15-18

Maynard Smith, J. (1974): The Theory of Games and the Evolution of Animal Conflicts. J. theor. Biology **47**, 209-221

Maynard Smith, J. (1982): Evolutionary Game Theory. Cambridge: Cambridge University Press

Mead, C. & L. Conway (1980): Introduction to VLSI Systems. Reading: Addison-Wesley

Mende, W., M. Peschel & H.-M. Voigt (1983): Einige Überlegungen zum vertieften dialektischen Verständnis von Evolution. Sitzungsberichte der Akademie der Wissenschaften der DDR, 10G/1983, 65-74. Berlin: Akademie-Verlag

Milne, L.J. & M. Milne (1960): The Balance of Nature. New York: Alfred A. Knopf

Mühlenbein, H., M. Gorges-Schleuter & O. Krämer (1987): New Solutions to the Mapping Problem of Parallel Systems. Parallel Computing 4, 269-279

Papadimetriou, C.H. & K. Steiglitz (1982): Combinatorial Optimization. Englewood-Cliffs: Prentice-Hall

Peschel, M. & W. Mende (1983): Leben wir in einer Volterra-Welt ? Ein ökologischer Zugang zur angewandten Systemanalyse. Berlin: Akademie- Verlag

Peschel, M. & W. Mende (1986): Predator-Prey-Models. Wien: Springer-Verlag & Berlin: Akademie-Verlag

Peschel, M., W. Mende & F. Breitenecker (1987): Anwendung der diskreten Theorie ratengekoppelter Systeme in der Regelungstechnik. messen - steuern - regeln **30**, 322-324

Peschel, M., H.-M. Voigt, W. Mende & F. Breitenecker (1988): System Engineering Methodology for Simulation and Control of Dynamical Networks. In: Proc. Int. Symp. Systems Analysis and Simulation. Berlin: Akademie-Verlag

Rechenberg, I. (1973): Evolutionsstrategie: Optimierung technischer Systeme nach Prinzipien der biologischen Evolution. Stuttgart: F. Frommann

Rechenberg, I. (1984): The Evolution Strategy - A Mathematical Model of Darwinian Evolution. In: Frehland, E: (Ed.): Springer Series in Synergetics **22**. Berlin, Heidelberg, New York: Springer Verlag, 122-132

Richardt, J. (1980): Eine unscharfe Strategie zur Lösung quadratische Zuordungsprobleme. Seminarberichte "Unscharfe Modellierung und Steuerung", TH Karl-Marx-Stadt, Teil III, 35-42

Richardt, J. (1983): Ein iteratives, auf Objekt-Klasse-Beziehungen basierendes Clusterverfahren und eine unscharfe Lösungsstrategie für Zuordnungsprobleme. Dissertation A, Akademie der Wissenschaften der DDR

Richardt, J. (1985): Lokale Ziele und die unscharfe "thermodynamische" Methode Ac. Sci. GDR, Inst. for Informatics, IIR-Report vol. 1, no. 12, 11-61

Savageau, M. (1976): Biochemical Systems Analysis. Reading: Addison-Wesley

Schuster, P., K. Sigmund, J. Hofbauer, R. Wolff, R. Gottlieb & P. Merz (1981): Selfregulation of Behaviour in Animal Societies. Biol. Cybernetics vol. **40**, 1-25

Schuster, P. (1986): Dynamics of Molecular Evolution. Physica **22D**, 100-119

Schuster, P. (1986): The Physical Basis of Molecular Evolution. Chemica Scripta **26B**, 27-41

Schwefel, H.P. (1977): Numerische Optimierung von Computermodellen mittels der Evolutionsstrategie. Basel, Stuttgart: Birkhäuser-Verlag

Scudo, F. & J. Ziegler (1978): The Golden Age of Theoretical Ecology 1923-1940.Lecture Notes in Biomathematics 22. Berlin, Heidelberg, New York: Springer-Verlag

Shahshahani, S. (1979): A New Mathematical Framework for the Study of Linkage and Selection. Memoirs AMS 211

Sigmund, K. (1985): The Maximum Principle for Replicator Equations. Proc. 5th Meeting UNESCO's Working Group on System Theory 1984, 63-71. Berlin: Akademie-Verlag

Svirezhev, Yu. M. & D.O. Logofet (1983): Stability of Biological Communities. Moscow: Mir Publ.

Tank, D.W. & J.J. Hopfield (1987): Collective Computation in Neuronlike Circuits. Scientific American, Dec. 1987, 62-70

Taylor, P. & L. Jonker (1978): Evolutionary Stable Strategies and Game Dynamics Math. Biosci. vol. 40, 145-156

v. Trzebiatowski, G. (1984): Über ein Verfahren zur Clusteranalyse und ihre Anwendung beim Layoutentwurf zur Partitionierung mikroelektronischer Schaltkreise. Dissertation A, Humboldt-Universität Berlin

v. Trzebiatowski, G. (1985): Thermodynamic Simulation Procedure for Partitioning Problems. Ac. Sci. GDR, Inst. for Informatics, IIR-Report vol. 1, no. 12, 86-96

Uehara, T. (1985): A Knowledge-Based Logic Design System. IEEE Design and Test, no. 10, 27-34

Ullman, J.D. (1984): Computational Aspects of VLSI. Potomac: Computer Science Press

Vecchi, M.P. & S. Kirkpatrick (1983): Global Wiring by Simulated Annealing. IEEE Trans. CAD, no. 2, 215-222

Verhulst, P.F. (1845): Recherches mathematique sur la taux d'accroisement de la population. Mem. Acad. Roy. Belgique 18, 1-38

Voigt, H.-M. (1977): DICSI: A Digital Circuit Simulation and Analysis Program. Proc. IFAC-Symp. on Discrete Systems, Dresden

Voigt, H.-M. & R. Witt (1977): DICSI: Anwendungsbeschreibung. Berlin: Inst. for Control Engineering

Voigt, H.-M. (1980): Generierung und Optimierung von Testfolgen für Digitalschaltungen im Mensch-Maschine-Dialog. Messen-Steuern-Regeln 23 no. 10, 569-579

Voigt, H.-M. (1982): Evolution Processes and Structuring. In: Proc. "Tagung numerische Lösung von Differentialgleichungen 1982". Ac. Sci. GDR, Inst. for Mathematics, Report R-MATH-01/83, 242-249

Voigt, H.-M. (1984): Structure Formation Processes, Vector Optimization, and Evolution: An Ecological Approach. Proc. 4th Joint Seminar "Nonconventional Optimization Methods", Warsaw, 26 pp.

Voigt, H.-M. (1985): Structure Formation, Vector Optimization, and Evolution. Proc. Int. Symp. Systems Analysis and Simulation, 251-254. Berlin: Akademie-Verlag

Voigt, H.-M. (1985): Structuring and Evolution Processes. Proc. 5th Meeting UNESCO's Working Group on System Theory 1984. Berlin: Akademie-Verlag

Voigt, H.-M. (1985): Structure Formation, Optimization, and Evolution. Wiss. Z. TH Ilmenau, no. 2, 117-132

Voigt, H.-M. (Ed.) (1985): Structure Formation Based on Self-Organization Principles with Applications to Electronics Design. Ac. Sci. GDR, Inst. for Informatics, IIR-Report vol. 1, no. 12

Voigt, H.-M., G. v. Trzebiatowski & J. Richardt (1985): A Comparative Study on Algorithms for the Partitioning of Graphs Based on Self-Organization Principles. Ac. Sci. GDR, Inst. for Informatics, IIR-Report vol. 1, no. 12, 62-85

Voigt, H.-M. (1986): Pattern Formation by Natural Selection with Applications to Decision Support. In: Proc. Int. Workshop on Large-Scale Modelling and Interactive Decision Analysis 1985. Berlin, Heidelberg, New York: Springer-Verlag

Voigt, H.-M. & I. Santibanez-Koref (1988): Solving Assignment Problems by Selection Pressure Controlled Replicator Networks. In: Proc. Int. Symp. Systems Analysis and Simulation. Berlin: Akademie-Verlag

Voigt, H.-M. & I. Santibanez-Koref (1988): Solving Complex Problems by Selection Pressure Controlled Replicator Networks. In: Proc. Int. Wiss. Kolloquium Ilmenau. Ilmenau: Technical University Press

Voigt, H.-M. & I. Santibanez-Koref (1988): Combinatorial Optimization by Selection Pressure Controlled Replicator Networks. In: Proc. IV. Int. Workshop on Parallel Processing by Cellular Automata and Arrays. Berlin: Akademie-Verlag

Voigt, H.-M. (1988): Optimization by Selection Pressure Controlled Replicator Networks. Systems Analysis, Modelling, and Simulation, vol. 5, no. 5

Volterra., V. (1931): Lecons sur la theorie mathematique de la lutte pour la vie. Paris: Gauthiers-Villars

Wilson, G.V. & G.S. Pawley (1987): On the Stability of the Travelling Salesman Algorithm of Hopfield and Tank. University of Edinburgh, Department of Physics

Wright, S. (1921): Systems of Mating. Genetics 6, 11-178

Wright, S. (1931): Evolution in Mendelian Populations. Genetics **16**, 97-159

Zeeman, E.C. (1980): Population Dynamics from Game Theory. In: Global Theory of Dynamical Systems. Lecture Notes in Mathematics **819**. Berlin, Heidelberg, New York: Springer-Verlag

Zeeman, E.C. (1981): Dynamics of Evolution of Animal Conflicts. J. theor. Biology vol. **89**, 249-270

Appendices

The Appendices contain three sample programs for the travelling salesman problem and for the module placement problem.

The progams are written in BASIC and run with QuickBasic, and TurboBasic as well, on IBM PC's and Compatibles with CGA graphics.

The functions of the travelling salesman program are activated by the following menu items:

GINP Graphic input of city positions on the screen.
For end of input move the marker to the upper-left corner of the screen and press the RETURN key.

NINP Numeric input of city positions (keyboard or random).
For end of input enter city position (0,0) and press the RETURN key.

I/O Storage and retrieval of city positions on disc.

DEF Change default values of selection pressure, selection pressure change, mutation rate, and storage device.

RUN Activate computation (iterations).
Interrupt the computation by pressing any key.

STOP Return to system.

LOW Decrease selection pressure.

HIGH Increase selection pressure.

TOUR Show salesman's tour on screen.

TEXT Put text on screen.

PICT Storage and retrieval of screen content.

HCPY Hardcopy of screen content to printer device.

ASSG Show assignments of cities to tour positions.

The menu items of the module placement program can be used in much the same way.

To use the marker on the screen press the NUM LOCK key and the keys on the numeric panel (up -> 8, down -> 2, right ->6, left -> 4).

Before starting press the CAPS LOCK key.

Dependent on the compiler you have to adjust the line width xxx in the WIDTH "LPT1:",xxx statement of the hardcopy subroutine.

Now you are ready to make your own experiences with replicator network algorithms. Proposals and comments with this respect are highly welcomed.

Good Luck !

Appendix A

Program for the travelling salesman problem

```
REM
REM
REM                    TRAVELLING SALESMAN
REM
REM          HANS-MICHAEL VOIGT     -    27.1.1988
REM
REM
REM INVERSE REPLICATOR EQUATION WITH SELECTION PRESSURE CONTROL
REM
REM       MUTATIONS WILL BE DONE FOR EVERY NEW GENERATION
REM
REM
```

```
REM ------------------------ MAIN PROGRAM ---------------------
 DIM G(50,50),X(50,50),F(50,50),XH(50),GZ(50)
 DIM PX(50),PY(50),XB(50),YB(50),M$(20)
 DIM N$(50),TA(50),MARK(40),MU(7),MO(7)
REM ----------------------- DEFAULT VALUES --------------------
 READ RC,FR,DC,N%,XM,YM,MX,MY,TRIALS%,DEV$
 FOR I%=1 TO 13
     READ M$(I%)
 NEXT I%
 FOR I%=0 TO 7
     READ MU(I%)
     READ MO(I%)
 NEXT I%
 DATA -.1,1.E-4,.01,50,0,0,304,3,0,A
 DATA GINP,NINP,I/O,DEF,RUN,STOP,LOW,HIGH
 DATA TOUR,TEXT,PICT,HCPY,ASSG
 DATA 0,4,0,3,0,3,0,4,3,5,4,6,5,7,6,6
 RANDOMIZE(VAL(RIGHT$(TIME$,2)))
 SCREEN 1:CLS
 COLOR 0,0
 COLOR 1,1
 GOSUB MARKER
 GOSUB UNIFORM
 GOSUB MENUE
M1:
 GOSUB SELECTMENUE
 ON MC GOSUB SGINP,SNINP,SIO,SDEF,SRUN,SSTOP,SLOW,SHIGH,STOUR,
             STEXT,SPICT,SHCPY,DRAWASSIGNMENTS
 GOTO M1
END
```

```
REM -------------------------- MENUE -------------------------
MENUE:
 FOR I%=1 TO 13
     LOCATE I%,35
     PRINT M$(I%)
 NEXT I%
RETURN
```

```
REM --------------------------- CHANGE DEFAULT VALUES ---------
SDEF:
 SCREEN 1:CLS
 PRINT "SELECTION PRESSURE =";RC;:INPUT VA:IF VA<>0 THEI RC=VA
 PRINT "MUTATION RATE      =";FR;:INPUT VA:IF VA<>0 THEN FR=VA
 PRINT "SEL. PRESS. CHANGE =";DC;:INPUT VA:IF VA<>0 THEN DC=VA
 PRINT "DEFAULT DEVICE     =";" "+DEV$;:INPUT R$:IF R$<>"" THEN DEV$=R$
 SCREEN 1:CLS
 GOSUB MENUE
RETURN
```

```
REM --------------------------- GRAPHIC INPUT -----------------
SGINP:
 XM=0
 YM=0
 X1=0
 Y1=0
 N%=-1
SGINP2:
 GOSUB MOVEMARKER
 IF X1=0 AND Y1=0 GOTO SGINP1
 N%=N%+1
 PX(N%)=X1
 PY(N%)=180-Y1
 XB(N%)=X1
 YB(N%)=Y1
 LINE (X1-1,Y1-1)-(X1+1,Y1+1),3,BF
 HX=INT((X1-1)/8)
 HY=INT((Y1-11)/8)
 GOTO SGINP2
SGINP1:
 GOSUB DISTANCES
 GOSUB NORMALIZATION
 GOSUB MUTATIONS
RETURN
```

```
REM ---------------------------- NUMERIC INPUT -----------------
SNINP:
 SCREEN 1:CLS
 IO$="KR"
 PRINT "KEYBOARD/RANDOM (K/R) ?"
NINP10:
 R$=INKEY$:IF R$="" GOTO NINP10
 H=INSTR(IO$,R$)
 IF H=0 GOTO NINP10
 ON H GOTO NINP2,NINP3
NINP2:
 N%=-1
NINP1:
 PRINT "NAME,X,Y = ";
 N%=N%+1
 INPUT N$(N%),PX(N%),PY(N%)
 IF PX(N%)<>0 OR PY(N%)<>0 GOTO NINP1
 N%=N%-1
 GOTO NINP4
NINP3:
 INPUT "# OF CITIES =";N%
 N%=N%-1
 INPUT "AREA LENGTH =";L
 FOR I%=0 TO N%
     PX(I%)=RND(1)*L
     PY(I%)=RND(1)*L
 NEXT I%
REM ------------------- SCALING -------------------------------
NINP4:
 XO=PX(0)
 XU=PX(0)
 YO=PY(0)
 YU=PY(0)
 FOR I%=1 TO N%
     IF PX(I%)>XO THEN XO=PX(I%)
     IF PX(I%)<XU THEN XU=PX(I%)
     IF PY(I%)>YO THEN YO=PY(I%)
     IF PY(I%)<YU THEN YU=PY(I%)
 NEXT I%
 HX=XO-XU
 HY=YO-YU
 FOR I%=0 TO N%
     XB(I%)=INT(150*(PX(I%)-XU)/HX)+10
     YB(I%)=INT(-150*(PY(I%)-YU)/HY)+170
 NEXT I%
 GOSUB DISTANCES
 GOSUB DRAWCITIES
 GOSUB NORMALIZATION
 GOSUB MUTATIONS
 GOSUB MENUE
RETURN
```

```
REM --------------------------- SAVE/LOAD DATA -----------------
SIO:
 SCREEN 1:CLS
 F$=DEV$+":TSP*"
 FILES DEV$+":*.*"
 IO$="LS"
 PRINT "LOAD/SAVE (L/S) ?"
IO1:
 R$=INKEY$:IF R$="" GOTO IO1 H=INSTR(IO$,R$)
 IF H=0 GOTO IO1
 PRINT "FILE NAME ="
 INPUT F$ ON H GOTO IO2,IO3
IO2:
 OPEN DEV$+":TSP"+F$ FOR INPUT AS #1
 INPUT#1,NO
 N%=NO
 FOR I%=0 TO N%
     INPUT#1,N$(I%)
     INPUT#1,PX(I%)
     INPUT#1,PY(I%)
     INPUT#1,XB(I%)
     INPUT#1,YB(I%)
 NEXT I%
 CLOSE #1
 CLS
 GOTO IO4
IO3:
 OPEN DEV$+":TSP"+F$ FOR OUTPUT AS #1
 NO=N%
 PRINT#1,NO
 FOR I%=0 TO N%
     PRINT#1,N$(I%)
     PRINT#1,PX(I%)
     PRINT#1,PY(I%)
     PRINT#1,XB(I%)
     PRINT#1,YB(I%)
 NEXT I%
 CLOSE #1
IO4:
 CLS
 GOSUB DISTANCES
 GOSUB DRAWCITIES
 GOSUB NORMALIZATION
 GOSUB MUTATIONS
 GOSUB MENUE
RETURN
```

```
REM ------------------------- COMPUTE ASSIGNMENTS -------------
SRUN:
 GOSUB NORMALIZATION
RUN2:
 TRIALS%=TRIALS%+1
 LOCATE 20,30
 PRINT TRIALS%;
 FOR K%=0 TO N%
     S=0
     FOR I%=0 TO N%
         S=S+X(I%,K%)
     NEXT I%
     XH(K%)=S
 NEXT K%
 FOR I%=0 TO N%
     FOR K%=0 TO N%
         S=0
         FOR J%=0 TO N%
             S=S+G(I%,J%)*X(J%,K%)
         NEXT J%
         F(I%,K%)=S
     NEXT K%
 NEXT I%
 IF RC<0 GOTO RUN10
 FOR K%=0 TO N%
     L%=K%-1
     R%=K%+1
     IF L%<0 THEN L%=L%+(N%+1)
     IF R%>N% THEN R%=R%-(N%+1)
     FOR I%=0 TO N%
         X(I%,K%)=
           X(I%,K%)*(1/(F(I%,L%)+F(I%,K%)+F(I%,R%))+RC*X(I%,K%))/XH(K%)
     NEXT I%
 NEXT K%
 GOTO RUN11
RUN10:
 FOR K%=0 TO N%
     L%=K%-1
     R%=K%+1
     IF L%<0 THEN L%=L%+(N%+1)
     IF R%>N% THEN R%=R%-(N%+1)
     FOR I%=0 TO N%
         X(I%,K%)=X(I%,K%)*
           (1/(F(I%,L%)+F(I%,K%)+F(I%,R%))-RC*(XH(K%)-X(I%,K%)))/XH(K%)
     NEXT I%
 NEXT K%
RUN11:
 GOSUB NORMALIZATION
 GOSUB MUTATIONS
 GOSUB STOUR
 R$=INKEY$:IF R$<>"" THEN RETURN
 GOTO RUN2
RETURN
```

```
REM ------------------------- EXIT TO SYSTEM -----------------
SSTOP:
 END
RETURN
```

```
REM ---------------------------- DECREMENT SELECTION PRESSURE --
SLOW:
 IF RC<= .01 THEN RC=RC-DC
 IF RC > .01 THEN RC=RC-DC/10
 GOSUB MUTATIONS
RETURN
```

```
REM ---------------------------- INCREMENT SELECTION PRESSURE --
SHIGH:
 IF RC>=-.01 THEN RC=RC+DC/10
 IF RC< -.01 THEN RC=RC+DC
 GOSUB MUTATIONS
RETURN
```

```
REM ---------------------------- COMPUTE TOUR ------------------
STOUR:
 SWITCH%=1
 FOR K%=0 TO N%
     S=0
     FOR I%=0 TO N%
         IF X(I%,K%)<=S GOTO TOUR1
         S=X(I%,K%)
         TA(K%)=I%
TOUR1:
     NEXT I%
     S=0
     FOR L%=0 TO N%
         IF X(TA(K%),L%)<=S GOTO TOUR3
         S=X(TA(K%),L%)
         HA%=L%
TOUR3:
     NEXT L%
     IF HA%=K% GOTO TOUR2
     TA(K%)=-1
     SWITCH%=0
TOUR2:
 NEXT K%
 SCREEN 1:CLS
 GOSUB MENUE
 GOSUB DRAWSTREETS
 GOSUB DRAWCITIES
REM -------------- COMPUTE TOUR LENGTH -----------------------
 IF SWITCH%=0 THEN RETURN
 TL=0
 FOR I%=0 TO N%-1
     TL=TL+G(TA(I%),TA(I%+1))
 NEXT I%
 TL=TL+G(TA(N%),TA(0))
 LOCATE 1,1
 PRINT "TOUR LENGTH ="+STR$(TL)
RETURN
```

```
REM ------------------------ OUTPUT TEXT ON GRAPHIC SCREEN -----
STEXT:
 GOSUB MOVEMARKER
 X1=INT(X1/8)
 Y1=INT(Y1/8)
TEXT1:
 R$=INKEY$:IF R$="" GOTO TEXT1
 IF R$=CHR$(13) THEN RETURN
 LOCATE Y1,X1
 PRINT R$
 X1=X1+1
 GOTO TEXT1
RETURN
```

```
REM ------------------------ SAVE/LOAD GRAPHICS ----------------
SPICT:
 F$=DEV$+":PICSCR.SCN"
 DEF SEG=&HB800
 BSAVE F$,0,&H4000
 SCREEN 1:CLS
 FILES DEV$+":*.*"
 IO$="LSK"
 PRINT "LOAD/SAVE/KILL (L/S/K) ?"
PICT1:
 R$=INKEY$:IF R$="" GOTO PICT1
 H=INSTR(IO$,R$)
 IF H=0 GOTO PICT1
 PRINT "FILE NAME ="
 F$=""
 INPUT F$
 F$=DEV$+":PIC"+F$+".SCN"
 ON H GOTO PICT2,PICT3,PICT4
PICT2:
 BLOAD F$
RETURN
PICT3:
 NAME DEV$+":PICSCR.SCN" AS F$
 CLS
 GOSUB MENUE
RETURN
PICT4:
 KILL F$
 CLS
 GOSUB MENUE
RETURN
```

```
REM ------------------------ HARDCOPY --------------------------
SHCPY:
 WIDTH "LPT1:",255
 LPRINT:LPRINT
 LPRINT CHR$(27);"A";
 LPRINT CHR$(8);
 DEF SEG=&HB800
 LOWADR=7919
 HIGHADR=-1
 FOR I=0 TO 79
     LOWADR=LOWADR+1
     HIGHADR=HIGHADR+1
     LOWADR1=LOWADR+8192
     LPRINT CHR$(27);"K";CHR$(144);CHR$(1);
     FOR J=LOWADR TO HIGHADR STEP -80
         LPRINT CHR$(PEEK(LOWADR1));CHR$(PEEK(LOWADR1));
                CHR$(PEEK(J));CHR$(PEEK(J));
         LOWADR1=LOWADR1-80
     NEXT J
 LPRINT
 NEXT I
 LPRINT CHR$(27);"@";
RETURN
```

```
REM ------------------------ DEFINE COLOURS --------------------
SCOL:
RETURN
```

```
REM ----------------------- UNIFORM DISTRIBUTION -------------
UNIFORM:
 FOR I%=0 TO N%
     FOR K%=0 TO N%
         X(I%,K%)=1
     NEXT K%
 NEXT I%
RETURN
```

```
REM ----------------------- MUTATIONS -----------------------
MUTATIONS:
 FOR I%=0 TO N%
     FOR K%=0 TO N%
         X(I%,K%)=X(I%,K%)+RND(1)*FR
     NEXT K%
 NEXT I%
 GOSUB NORMALIZATION
RETURN
```

```
REM ----------------------- NORMALIZATION --------------------
NORMALIZATION:
 FOR I%=0 TO N%
     S=0
     FOR K%=0 TO N%
         S=S+X(I%,K%)
     NEXT K%
     FOR K%=0 TO N%
         X(I%,K%)=X(I%,K%)/S
     NEXT K%
 NEXT I%
RETURN
```

```
REM ----------------------- DISTANCES -----------------------
DISTANCES:
 SCREEN 1:CLS
 PRINT "EUCLIDEAN/MANHATTAN DISTANCE ( E/M ) ?";
 B$="EM"
DIST1:
 R$=INKEY$:IF R$="" GOTO DIST1
 H=INSTR(B$,R$)
 IF H=0 GOTO DIST1
 SCREEN 1
 ON H GOTO DIST2,DIST3
DIST2:
 FOR I%=0 TO N%-1
     FOR J%=I%+1 TO N%
         HX=(PX(I%)-PX(J%))^2
         HY=(PY(I%)-PY(J%))^2
         G(I%,J%)=(SQR(HX+HY))
         G(J%,I%)=G(I%,J%)
     NEXT J%
 NEXT I%
 GOTO DIST4
DIST3:
 FOR I%=0 TO N%-1
     FOR J%=I%+1 TO N%
         HX=ABS(PX(I%)-PX(J%))
         HY=ABS(PY(I%)-PY(J%))
         G(I%,J%)=((HX+HY))
         G(J%,I%)=G(I%,J%)
     NEXT J%
 NEXT I%
DIST4:
 SCREEN 1:CLS
 GOSUB MENUE
RETURN
```

```
REM ------------------------ MOVE MARKER ---------------------
MOVEMARKER:
 X1=XM
 Y1=YM
 PUT (XM,YM),MARK,XOR
 C$=CHR$(52)+CHR$(54)+CHR$(56)+CHR$(50)+CHR$(13)
MOVE1:
 R$=INKEY$:IF R$="" GOTO MOVE1
 H=INSTR(C$,R$)
 IF H=0 GOTO MOVE1
 ON H GOTO MOVE2,MOVE3,MOVE4,MOVE5,MOVE6
MOVE2:
 IF XM<=0 GOTO MOVE1
 PUT (XM,YM),MARK,XOR
 XM=XM-5
 X1=X1-5
 GOTO MOVE7
MOVE3:
 IF XM>=310 GOTO MOVE1
 PUT (XM,YM),MARK,XOR
 XM=XM+5
 X1=X1+5 GOTO MOVE7
MOVE4:
 IF YM<=0 GOTO MOVE1
 PUT (XM,YM),MARK,XOR
 YM=YM-5
 Y1=Y1-5
 GOTO MOVE7
MOVE5:
 IF YM>=190 GOTO MOVE1
 PUT (XM,YM),MARK,XOR
 YM=YM+5
 Y1=Y1+5
MOVE7:
 PUT (XM,YM),MARK,XOR GOTO MOVE1
MOVE6:
 PUT (XM,YM),MARK,XOR
RETURN
```

```
REM -------------------------- DEFINE MARKER -----------------
MARKER:
 FOR I%=0 TO 7
     FOR J%=MU(I%) TO MO(I%)
         PSET(J%,I%),3
     NEXT J%
 NEXT I%
 GET (0,0)-(7,7),MARK
 CLS
RETURN
```

```
REM -------------------------- SELECT MENUE -------------------
SELECTMENUE:
 PUT (MX,MY),MARK,XOR
 C$=CHR$(56)+CHR$(50)+CHR$(13)
MARKER1:
 R$=INKEY$:IF R$="" GOTO MARKER1
 H=INSTR(C$,R$)
 IF H=0 GOTO MARKER1
 ON H GOTO MARKER2,MARKER3,MARKER4
MARKER2:
 IF MY<=3 GOTO MARKER1
 PUT (MX,MY),MARK,XOR
 MY=MY-8
 GOTO MARKER5
MARKER3:
 IF MY>=99
 GOTO MARKER1
 PUT (MX,MY),MARK,XOR
 MY=MY+8
MARKER5:
 PUT (MX,MY),MARK,XOR
 GOTO MARKER1
MARKER4:
 PUT (MX,MY),MARK,XOR
 MC=(MY-3)/8+1
RETURN
```

```
REM ---------------------------- DRAW ASSIGNMENTS ---------------
DRAWASSIGNMENTS:
 SCREEN 1:CLS
 GOSUB MENUE
 FOR I%=0 TO N%
     FOR K%=0 TO N%
          S=INT(X(I%,K%)*8)
          X1=20 +K%*8-INT(X(I%,K%)*4)
          Y1=20 +I%*8-INT(X(I%,K%)*4)
          LINE (X1,Y1)-(X1+S,Y1+S),3,BF
     NEXT K%
 NEXT I%
 LOCATE 20,30
 PRINT TRIALS%;
RETURN
```

```
REM -------------------------- DRAW CITIES --------------------
DRAWCITIES:
 FOR I%=0 TO N%
     LINE (XB(I%)-1,YB(I%)-1)-(XB(I%)+1,YB(I%)+1),3,BF
     REM  LOCATE YB(I%)/8+2,XB(I%)/8+2
     REM  PRINT N$(I%);
 NEXT I%
RETURN
```

```
REM -------------------------- DRAW STREETS --------------------
DRAWSTREETS:
 FOR I%=0 TO N%-1
     IF TA(I%)=-1 OR TA(I%+1)=-1 GOTO DRAW1
     LINE (XB(TA(I%)),YB(TA(I%)))-(XB(TA(I%+1)),YB(TA(I%+1))),3
DRAW1:
 NEXT I%
 IF TA(0)=-1 OR TA(N%)=-1 GOTO DRAW2
 LINE (XB(TA(0)),YB(TA(0)))-(XB(TA(N%)),YB(TA(N%))),3
DRAW2:
RETURN
```

Appendix B

Program for the module placement problem

```
REM
REM
REM                        MODULE PLACEMENT
REM
REM          HANS-MICHAEL VOIGT    -   27.1.1988
REM
REM
REM    REPLICATOR EQUATION WITH SELECTION PRESSURE CONTROL
REM
REM      MUTATIONS WILL BE DONE FOR EVERY NEW GENERATION
REM
REM
```

```
REM ------------------------------ MAIN PROGRAM ----------------
 DIM G(50,50),X(50,3,12),F(50,3,12),XH(3,12)
 DIM M$(20),PROW%(50),PCOL%(50)
 DIM MARK(40),MU(7),MO(7)
REM ------------------------------ DEFAULT VALUES --------------
 READ RC,FR,DC,N%,XM,YM,MX,MY,TRIALS%,DEV$,ROW%,COL%
 FOR I%=1 TO 12
     READ M$(I%)
 NEXT I%
 FOR I%=0 TO 7
     READ MU(I%)
     READ MO(I%)
 NEXT I%
 DATA -.1,1.E-4,.01,50,0,0,304,3,0,A,3,12
 DATA INP,I/O,DEF,RUN,STOP,LOW,HIGH
 DATA TEXT,PICT,HCPY,ASSG,PLCE
 DATA 0,4,0,3,0,3,0,4,3,5,4,6,5,7,6,6
 RANDOMIZE(VAL(RIGHT$(TIME$,2)))
 SCREEN 1:CLS
 COLOR 0,0
 COLOR 1,1
 GOSUB MARKER
 GOSUB UNIFORM
 GOSUB MENUE
M1:
 GOSUB SELECTMENUE
 ON MC GOSUB SNINP,SIO,SDEF,SRUN,SSTOP,SLOW,SHIGH,STEXT,
             SPICT,SHCPY,DRAWASSIGNMENTS,DRAWPLACE
 GOTO M1
END
```

```
REM -------------------------- MENUE --------------------------
MENUE:
 FOR I%=1 TO 12
     LOCATE I%,35
     PRINT M$(I%)
 NEXT I%
RETURN
```

```
REM --------------------------- CHANGE DEFAULT VALUES ----------
SDEF:
 SCREEN 1:CLS
 ROW%=ROW%+1
 COL%=COL%+1
 PRINT "# OF ROWS           =";ROW%;:INPUT VA:IF VA<>0 THEN ROW%=VA
 PRINT "# OF COLUMNS        =";COL%;:INPUT VA:IF VA<>0 THEN COL%=VA
 PRINT "SELECTION PRESSURE =";RC;:INPUT VA:IF VA<>0 THEN RC=VA
 PRINT "MUTATION RATE      =";FR;:INPUT VA:IF VA<>0 THEN FR=VA
 PRINT "SEL. PRESS. CHANGE =";DC;:INPUT VA:IF VA<>0 THEN DC=VA
 PRINT "DEFAULT DEVICE     =";" "+DEV$;:INPUT R$:IF R$<>"" THEN DEV$=R$
 ROW%=ROW%-1
 COL%=COL%-1
 SCREEN 1:CLS
 GOSUB MENUE
RETURN
```

```
REM --------------------------- NUMERIC INPUT ------------------
SNINP:
 SCREEN 1:CLS
 PRINT "# OF MODULES =";:INPUT N%
 PRINT "# OF ROWS    =";:INPUT ROW%
 PRINT "# OF COLUMNS =";:INPUT COL%
 N%=N%-1
 ROW%=ROW%-1
 COL%=COL%-1
 FOR I%=0 TO N%-1
     FOR J%=I%+1 TO N%
         PRINT "G("+STR$(I%+1)+","+STR$(J%+1)+") =";:INPUT G(I%,J%)
         G(J%,I%)=G(I%,J%)
     NEXT J%
 NEXT I%
 GOSUB NORMALIZATION
 GOSUB MUTATIONS
 GOSUB MENUE
RETURN
```

```
REM -------------------------- SAVE/LOAD DATA ------------------
SIO:
 SCREEN 1:CLS
 F$=DEV$+":*"
 FILES DEV$+":*.*"
 IO$="LS"
 PRINT "LOAD/SAVE (L/S) ?"
IO1:
 R$=INKEY$:IF R$="" GOTO IO1
 H=INSTR(IO$,R$)
 IF H=0 GOTO IO1
 PRINT "FILE NAME =" INPUT F$
 ON H GOTO IO2,IO3
IO2:
 OPEN DEV$+":"+F$ FOR INPUT AS #1
 INPUT#1,NO
 N%=NO
 FOR I%=0 TO N%
     FOR J%=0 TO N%
         INPUT#1,G(I%,J%)
     NEXT J%
 NEXT I%
 CLOSE #1
 CLS
 PRINT "# OF MODULS  =";STR$(N%+1)
 PRINT "# OF ROWS    =";:INPUT ROW%
 PRINT "# OF COLUMNS =";:INPUT COL%
 ROW%=ROW%-1
 COL%=COL%-1
 CLS
 GOTO IO4
IO3:
 OPEN DEV$+":"+F$ FOR OUTPUT AS #1
 NO=N%
 PRINT#1,NO
 FOR I%=0 TO N%
     FOR J%=0 TO N%
         PRINT#1,G(I%,J%)
     NEXT J%
 NEXT I%
 CLOSE #1
IO4:
 CLS
 GOSUB NORMALIZATION
 GOSUB MUTATIONS
 GOSUB MENUE
RETURN
```

```
REM -------------------------- COMPUTE ASSIGNMENTS --------------
SRUN:
 GOSUB NORMALIZATION
RUN2:
 TRIALS%=TRIALS%+1
 LOCATE 20,30
 PRINT TRIALS%;
 FOR K%=0 TO ROW%
     FOR L%=0 TO COL%
          S=0
          FOR I%=0 TO N%
               S=S+X(I%,K%,L%)
          NEXT I%
          XH(K%,L%)=S
     NEXT L%
 NEXT K%
 FOR I%=0 TO N%
     FOR K%=0 TO ROW%
          FOR L%=0 TO COL%
               S=0
               FOR J%=0 TO N%
                    IF G(I%,J%)=0 GOTO RUN10
                    H=0
                    IF K%>0     THEN H=H+X(J%,K%-1,L%)
                    IF K%<ROW%  THEN H=H+X(J%,K%+1,L%)
                    IF L%>0     THEN H=H+X(J%,K%,L%-1)
                    IF L%<COL%  THEN H=H+X(J%,K%,L%+1)
                    S=S+G(I%,J%)*H
RUN10:
               NEXT J%
               F(I%,K%,L%)=S
          NEXT L%
     NEXT K%
 NEXT I%
 IF RC>=0 GOTO RUN20
 FOR I%=0 TO N%
     FOR K%=0 TO ROW%
          FOR L%=0 TO COL%
               X(I%,K%,L%)=X(I%,K%,L%)*
                    (F(I%,K%,L%)-RC*(1-X(I%,K%,L%)))/XH(K%,L%)
          NEXT L%
     NEXT K%
 NEXT I%
 GOTO RUN30
RUN20:
 FOR I%=0 TO N%
     FOR K%=0 TO ROW%
          FOR L%=0 TO COL%
               X(I%,K%,L%)=X(I%,K%,L%)*
                    (F(I%,K%,L%)+RC*X(I%,K%,L%))/XH(K%,L%)
          NEXT L%
     NEXT K%
 NEXT I%
RUN30:
 GOSUB NORMALIZATION
 GOSUB MUTATIONS
 GOSUB DRAWASSIGNMENTS
 R$=INKEY$:IF R$<>"" THEN RETURN
 GOTO RUN2
RETURN
```

```
REM -------------------------- EXIT TO SYSTEM ------------------
SSTOP:
 END
RETURN
```

```
REM ---------------------------- DECREMENT SELECTION PRESSURE ---
SLOW:
 IF RC<= .01 THEN RC=RC-DC
 IF RC > .01 THEN RC=RC-DC/10
 GOSUB MUTATIONS
RETURN
```

```
REM ------------------------ INCREMENT SELECTION PRESSURE ------
SHIGH:
 IF RC>=-.01 THEN RC=RC+DC/10
 IF RC< -.01 THEN RC=RC+DC
 GOSUB MUTATIONS
RETURN
```

```
REM ----------------------- OUTPUT TEXT ON GRAPHIC SCREEN ------
STEXT:
 GOSUB MOVEMARKER
 X1=INT(X1/8)
 Y1=INT(Y1/8)
TEXT1:
 R$=INKEY$:IF R$="" GOTO TEXT1
 IF R$=CHR$(13) THEN RETURN
 LOCATE Y1,X1
 PRINT R$
 X1=X1+1
 GOTO TEXT1
RETURN
```

```
REM ---------------------- SAVE/LOAD GRAPHICS -----------------
SPICT:
 F$=DEV$+":PICSCR.SCN"
 DEF SEG=&HB800
 BSAVE F$,0,&H4000
 SCREEN 1:CLS
 FILES DEV$+":*.*"
 IO$="LSK"
 PRINT "LOAD/SAVE/KILL (L/S/K) ?"
PICT1:
 R$=INKEY$:IF R$="" GOTO PICT1
 H=INSTR(IO$,R$)
 IF H=0 GOTO PICT1
 PRINT "FILE NAME ="
 F$=""
 INPUT F$
 F$=DEV$+":PIC"+F$+".SCN"
 ON H GOTO PICT2,PICT3,PICT4
PICT2:
 BLOAD F$
RETURN
PICT3:
 NAME DEV$+":PICSCR.SCN" AS F$
 CLS
 GOSUB MENUE
RETURN
PICT4:
 KILL F$
 CLS
 GOSUB MENUE
RETURN
```

```
REM ----------------------- HARDCOPY ----------------------------
SHCPY:
 WIDTH "LPT1:",500
 LPRINT:LPRINT
 LPRINT CHR$(27);"A";
 LPRINT CHR$(8);
 DEF SEG=&HB800
 LOWADR=7919
 HIGHADR=-1
 FOR I=0 TO 79
     LOWADR=LOWADR+1
     HIGHADR=HIGHADR+1
     LOWADR1=LOWADR+8192
     LPRINT CHR$(27);"K";CHR$(144);CHR$(1);
     FOR J=LOWADR TO HIGHADR STEP -80
         LPRINT CHR$(PEEK(LOWADR1));CHR$(PEEK(LOWADR1));
                CHR$(PEEK(J));CHR$(PEEK(J));
         LOWADR1=LOWADR1-80
     NEXT J
     LPRINT
 NEXT I
 LPRINT CHR$(27);"@";
RETURN
```

```
REM ----------------------- DEFINE COLOURS ---------------------
SCOL:
RETURN
```

```
REM --------------------- UNIFORM DISTRIBUTION --------------------
UNIFORM:
 FOR I%=0 TO N%
     FOR K%=0 TO ROW%
         FOR L%=0 TO COL%
             X(I%,K%,L%)=1
         NEXT L%
     NEXT K%
 NEXT I%
RETURN
```

```
REM ------------------- MUTATIONS -------------------------------
MUTATIONS:
 FOR I%=0 TO N%
     FOR K%=0 TO ROW%
         FOR L%=0 TO COL%
             X(I%,K%,L%)=X(I%,K%,L%)+RND(1)*FR
         NEXT L%
     NEXT K%
 NEXT I%
 GOSUB NORMALIZATION
RETURN
```

```
REM --------------- NORMALIZATION -------------------------------
NORMALIZATION:
 FOR I%=0 TO N%
     S=0
     FOR K%=0 TO ROW%
          FOR L%=0 TO COL%
               S=S+X(I%,K%,L%)
          NEXT L%
     NEXT K%
     FOR K%=0 TO ROW%
          FOR L%=0 TO COL%
               X(I%,K%,L%)=X(I%,K%,L%)/S
          NEXT L%
     NEXT K%
 NEXT I%
RETURN
```

```
REM ------------------- MOVE MARKER -----------------------------
MOVEMARKER:
 X1=XM
 Y1=YM
 PUT (XM,YM),MARK,XOR
 C$=CHR$(52)+CHR$(54)+CHR$(56)+CHR$(50)+CHR$(13)
MOVE1:
 R$=INKEY$:IF R$="" GOTO MOVE1
 H=INSTR(C$,R$)
 IF H=0 GOTO MOVE1
 ON H GOTO MOVE2,MOVE3,MOVE4,MOVE5,MOVE6
MOVE2:
 IF XM<=0 GOTO MOVE1
 PUT (XM,YM),MARK,XOR
 XM=XM-5
 X1=X1-5
 GOTO MOVE7
MOVE3:
 IF XM>=310 GOTO MOVE1
 PUT (XM,YM),MARK,XOR
 XM=XM+5
 X1=X1+5
 GOTO MOVE7
MOVE4:
 IF YM<=0 GOTO MOVE1
 PUT (XM,YM),MARK,XOR
 YM=YM-5
 Y1=Y1-5
 GOTO MOVE7
MOVE5:
 IF YM>=190 GOTO MOVE1
 PUT (XM,YM),MARK,XOR
 YM=YM+5
 Y1=Y1+5
MOVE7:
 PUT (XM,YM),MARK,XOR
 GOTO MOVE1
MOVE6:
 PUT (XM,YM),MARK,XOR
RETURN
```

```
REM ------------------------- DEFINE MARKER ---------------------
MARKER:
 FOR I%=0 TO 7
     FOR J%=MU(I%) TO MO(I%)
         PSET(J%,I%),3
     NEXT J%
 NEXT I%
 GET (0,0)-(7,7),MARK
 CLS
RETURN
```

```
REM -------------------- SELECT MENUE --------------------------
SELECTMENUE:
 PUT (MX,MY),MARK,XOR
 C$=CHR$(56)+CHR$(50)+CHR$(13)
MARKER1:
 R$=INKEY$:IF R$="" GOTO MARKER1
 H=INSTR(C$,R$)
 IF H=0 GOTO MARKER1
 ON H GOTO MARKER2,MARKER3,MARKER4
MARKER2:
 IF MY<=3 GOTO MARKER1
 PUT (MX,MY),MARK,XOR
 MY=MY-8
 GOTO MARKER5
MARKER3:
 IF MY>=91
 GOTO MARKER1
 PUT (MX,MY),MARK,XOR
 MY=MY+8
MARKER5:
 PUT (MX,MY),MARK,XOR
 GOTO MARKER1
MARKER4:
 PUT (MX,MY),MARK,XOR
 MC=(MY-3)/8+1
RETURN
```

```
REM -------------------- DRAW ASSIGNMENTS -----------------------
DRAWASSIGNMENTS:
 SCREEN 1:CLS
 GOSUB MENUE
 FOR K%=0 TO ROW%
     FOR L%=0 TO COL%
         XA=10+L%*(N%*3+12)
         YA=10+K%*60
         FOR I%=0 TO N%
             XXX=50*X(I%,K%,L%)
             LINE (XA,YA)-(XA+1,YA+XXX),3,BF
             XA=XA+3
         NEXT I%
     NEXT L%
 NEXT K%
 LOCATE 20,30
 PRINT TRIALS%;
RETURN
```

```
REM -------------------- DRAW PLACEMENT -------------------------
DRAWPLACE:
 SCREEN 1:CLS
 GOSUB MENUE
 FOR I%=0 TO N%
     MAX%=0
     FOR K%=0 TO ROW%
         FOR L%=0 TO COL%
             IF X(I%,K%,L%)<=MAX% GOTO PLACE10
             MAX%=X(I%,K%,L%)
             PROW%(I%)=K%
             PCOL%(I%)=L%
PLACE10:
         NEXT L%
     NEXT K%
 NEXT I%
 FOR I%=0 TO N%-1
     K%=PROW%(I%)
     L%=PCOL%(I%)
     FOR J%=I%+1 TO N%
         IF G(I%,J%)=0 GOTO PLACE20
         KK%=PROW%(J%)
         LL%=PCOL%(J%)
PLACE20:
     NEXT J%
 NEXT I%
 FOR K%=0 TO ROW%
     YA=K%*40+10
     FOR L%=0 TO COL%
         XA=L%*40+10
         LINE (XA,YA)-(XA+20,YA+20),3,BF
         LOCATE YA/8,XA/8
         MAX=0
         FOR I%=0 TO N%
             IF X(I%,K%,L%)<=MAX GOTO PLACE50
             MAX=X(I%,K%,L%)
             II%=I%
PLACE50:
         NEXT I%
         PRINT II%+1;
     NEXT L%
 NEXT K%
RETURN
```

Appendix C

Multilayer Replicator Network Algorithms

Within Sections 4. and 5. we developed the fundamentals for replicator network algorithms. We showed the application for the graph partitioning problem, for the module placement problem, and for the travelling salesman problem.

The algorithms are based on the diversified replicator equation as a unified approach. But between the graph partitioning algorithm and the travelling salesman algorithm as well as the module placement algorithm exists a decisive difference. For graph partitioning a super vertex (partition) may finally contain more then one vertex of the original graph. This is not the case for both the module placement problem and the travelling salesman problem. A super vertex (position of salesman s tour, place of a module) has to contain finally only one vertex of the original graph. This is a very strong constraint which may lead despite of the global influence of the selection pressure control to solutions which are far from the global best solution.

From a mathematical point of view, the derived algorithms do not take into account the internal structure of the given data.

Hence we raise the question how nature treats this problem. Though this question cannot be answered in a general way we want to show some typical examples leading to multilayer replicator network algorithms.

(i) In ecology a biotope is always structured as a multilayer system representing as a whole a food web. Here, a layer contains species which depend on the next lower layer and which are dependent on the next higher layer.

(ii) The rank order within highly organized animal societies can be seen as a multilayer system where a layer corresponds to a certain type of behaviour.

(iii) Perhaps the most prominent example of a multilayer system is the human brain. The system for visual perception consists of the eye and the optical tract which are connected via the thalamus and the reticular complex to the cortex. The striate cortex itself is horizontally structured into six layers. The vertical structure is given by macro-columns and columns.

Having this considerations in mind we may at once deduce multilayer replicator network algorithms in the following way:

Let us assume that the structure of the super graph corresponding to Sections 4. and 5. is given for the problem under consideration. In case of the travelling salesman problem the structure is given by a circuit. Then it is possible to use at a first layer much less super vertices then it is necessary for the original problem. Traversing top down to the second layer we may split each super vertex from the first layer into super vertices of the second layer. This splitting can be repeated until we get the original structure of the super graph. The transition from lower level layers to higher level layers corresponds now to a joining of lower level super vertices to higher level super vertices. At every layer the diversified replicator equation with equal fitness functions including a selection pressure control will be used for the dynamic unfolding of the solution.

The principal multilayer structure of a super vertex is shown in Figure C.1.

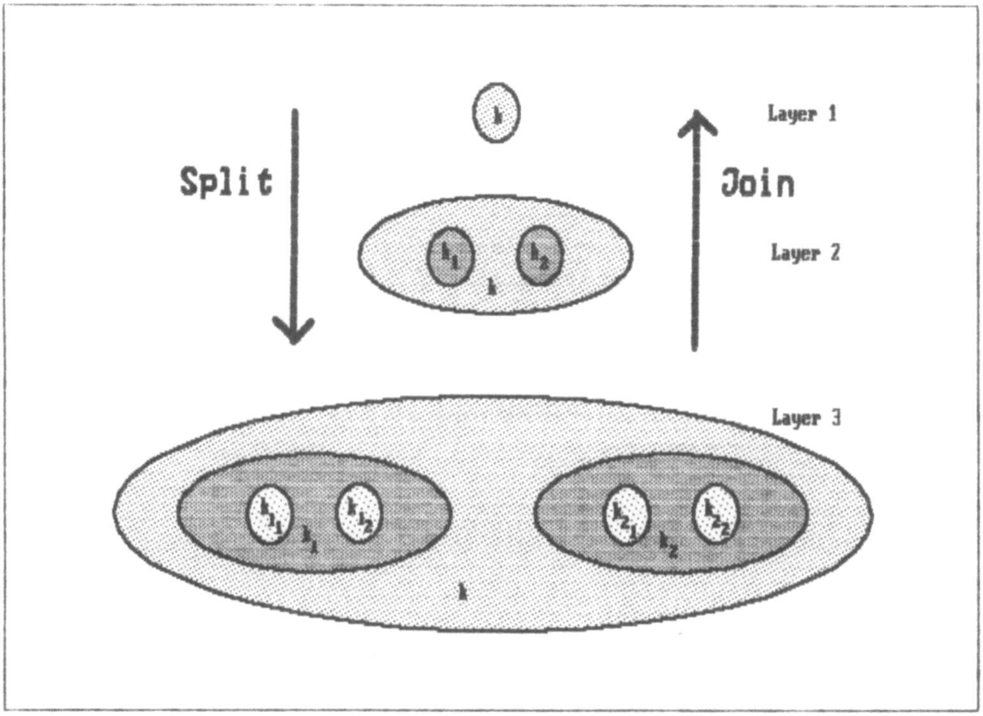

Figure C.1 Principal multilayer structure of a super vertex

Doing a split or a join operation one has to observe the following relations for the probabilities at every layer

$$\text{Layer 1}\quad x_{ik} \geq 0 \quad : \quad \sum_i x_{ik} = X_k$$

$$\text{Split} \downarrow \qquad \uparrow \text{Join}$$

$$\text{Layer 2}\quad x_{ik_l} \geq 0 \quad : \quad \sum_i x_{ik_l} = X_{k_l} \quad , \quad \sum_l X_{k_l} = X_k$$

$$\text{Split} \downarrow \qquad \uparrow \text{Join}$$

$$\text{Layer 3}\quad x_{ik_{l_r}} \geq 0 \quad : \quad \sum_i x_{ik_{l_r}} = X_{k_{l_r}} \quad , \quad \sum_r X_{k_{l_r}} = X_{k_l}$$

$$\text{Split} \downarrow \qquad \uparrow \text{Join}$$

. . .

Splitting and joining can be done for every iteration of the diversified replicator equation or only for specified overall system states.

We show the application of multilayer replicator network algorithms by means of the travelling salesman problem. For this problem we used only the split operation. Starting with a small number of super vertices (i.e. with a small number of city clusters) at the first layer an evolution process corresponding to Section 5. leads to the assignment of cities to city clusters. If the assignments are sufficiently crisp the city clusters will be splitted into smaller city clusters of the second layer. Probabilities which are smaller than a certain threshold will be set to zero. Then a new evolution process will be started. This procedure has to be repeated from layer to layer until the final tour will be yielded.

This multilayer replicator network algorithm shows very good results.

We want to verify its performance and efficiency by means of the Hopfield & Tank 30 city example. The city positions for this example were estimated from Figure 4 of Hopfield & Tank (1985) and are presented in Table C.1.

city #	x-position	y-position	city #	x-position	y-position
1	.071	.820	16	.194	.460
2	.361	.907	17	.348	.447
3	.458	.839	18	.219	.416
4	.677	.839	19	.929	.422
5	.748	.863	20	.826	.379
6	.761	.739	21	.606	.261
7	.703	.714	22	.368	.267
8	.574	.696	23	.413	.242
9	.426	.706	24	.381	.193
10	.316	.627	25	.594	.224
11	.587	.646	26	.535	.186
12	.671	.590	27	.387	.124
13	.665	.553	28	.490	.043
14	.606	.547	29	.987	.019
15	.355	.584	30	.626	.075

Table C.1 City positions of the Hopfield & Tank 30 city example

For this example we computed 30 runs on an IBM PC Compatible with different cluster sizes, different selection pressure schedules, and different threshold values. The results are shown in Table C.2.

	minimum	average	maximum
Tour length	4.284380	-.314401	4.362669
Iterations	140	332	622

Table C.2 Results for 30 runs with the multilayer replicator network algorithm for the Hopfield & Tank 30 city example

The best solution with the Lin & Kernighan algorithm yielded a tour length of 4.283504 and is depicted in Figure C.2.

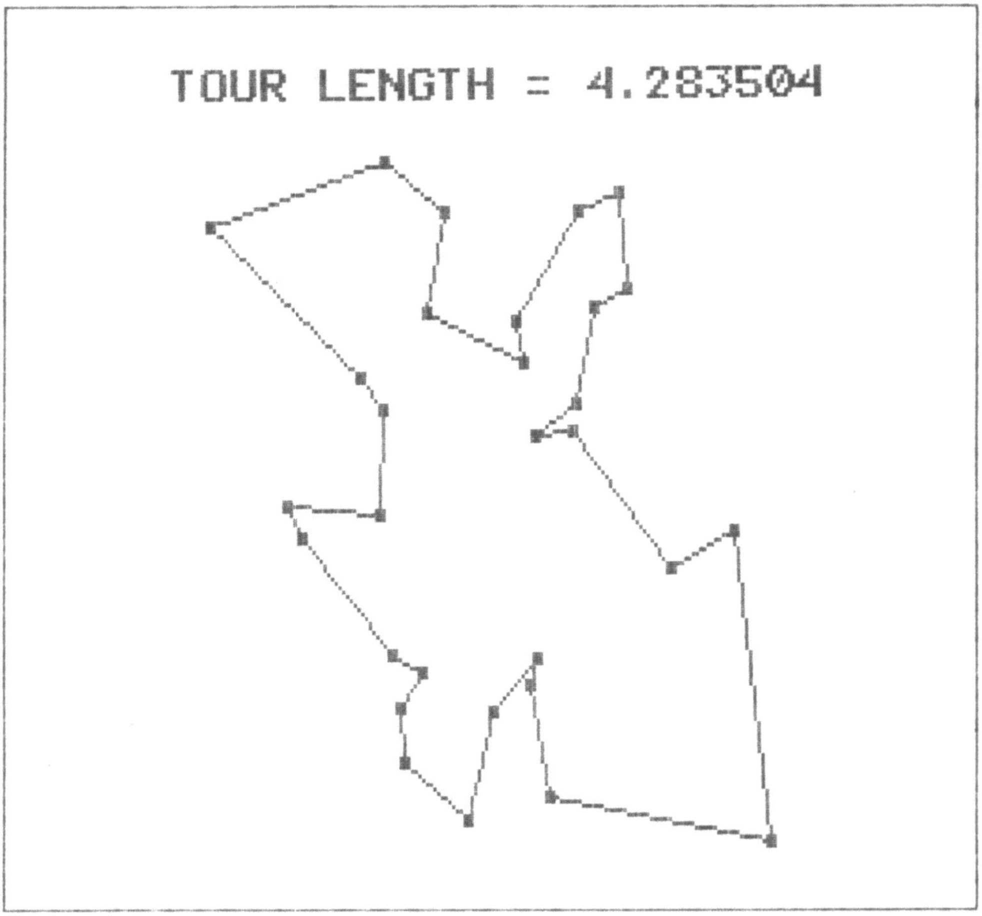

Figure C.2 Best solution for the Hopfield & Tank 30 city example with the Lin & Kernighan algorithm

The best solution with the multilayer replicator network algorithm was obtained with 3 layers where layer 1 contained 5 city clusters (super vertices), layer 2 contained 10 city clusters, and layer 3 contained 30 city clusters. The threshold was set to 0.1. The tour is only 0.02 % longer then the best Lin & Kernighan tour. The assignment of cities to city clusters for the different layers are shown in Figure C.3, and the final tour is depicted in Figure C.4.

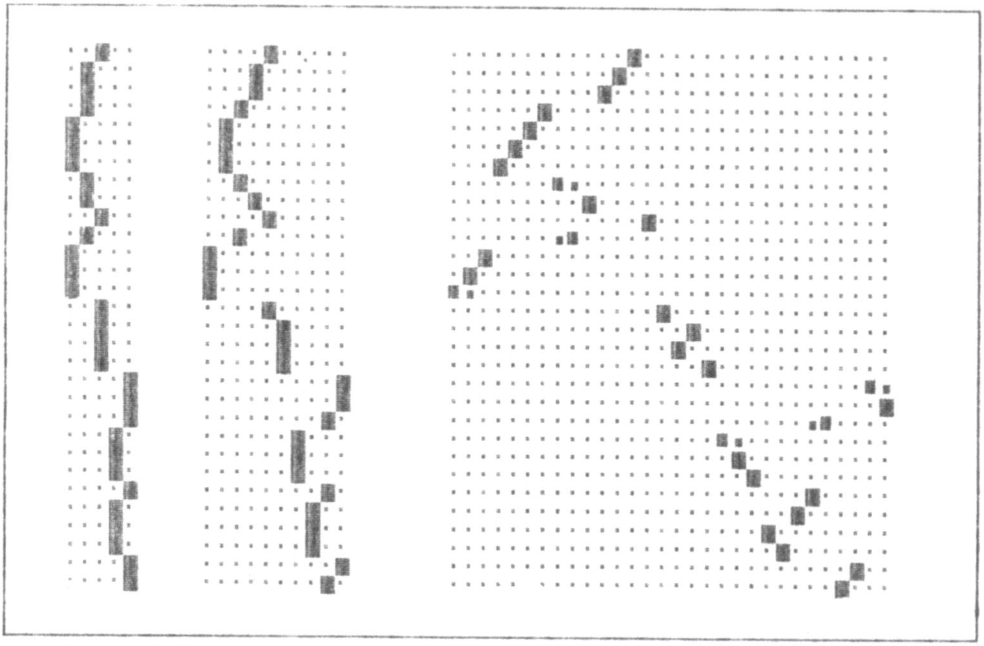

Figure C.3 Assignment of cities (rows) to city clusters (columns) on the first, second, and third layer for the Hopfield & Tank 30 city example

To run this algorithm on your PC you have to replace the program modules of Appendix A by the following modules. Furthermore you have to add the module for super vertex splitting. For this purpose the menu item

SPLT	Split a super vertex (city cluster) into a specified number of super vertices (city clusters)

was included.

It is very interesting that the Kolmogoroff entropy of the algorithm, i.e. the program length, it not increased significantly.

You should try to rewrite the program using sparse matrix techniques thus saving computer time and memory !

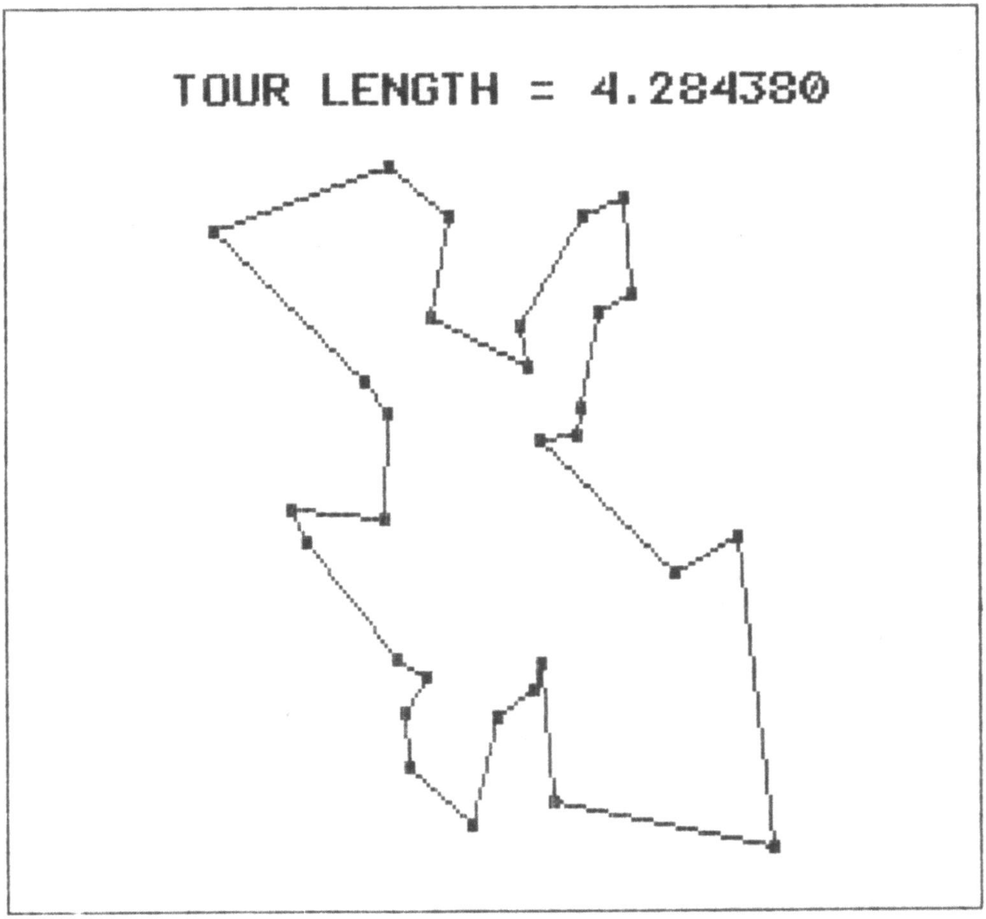

Figure C.4 Final best tour for the Hopfield & Tank 30 city example with the multilayer replicator network algorithm

```
REM
REM
REM                    TRAVELLING SALESMAN
REM
REM             HANS-MICHAEL VOIGT    -   13.6.1988
REM
REM
REM      INVERSE REPLICATOR EQUATION WITH SELECTION PRESSURE CONTROL
REM
REM           MUTATIONS WILL BE DONE FOR EVERY NEW GENERATION
REM
REM               MULTILAYER REPLICATOR NETWORK ALGORITHM
REM
REM
```

```
REM ---------------------------- MAIN PROGRAM ----------------
 DIM G(50,50),X(50,50),F(50,50),XH(50)
 DIM PX(50),PY(50),XB(50),YB(50),M$(20)
 DIM N$(50),TA(50),MARK(40),MU(7),MO(7)
REM ---------------------------- DEFAULT VALUES --------------
 READ RC,FR,TH,DC,N%,NC%,XM,YM,MX,MY,TRIALS%,DEV$
 FOR I%=1 TO 14
     READ M$(I%)
 NEXT I%
 FOR I%=0 TO 7
     READ MU(I%)
     READ MO(I%)
 NEXT I%
 DATA -.1,1.E-4,.05,.01,50,9,0,0,304,3,0,A
 DATA GINP,NINP,I/O,DEF,RUN,STOP,LOW,HIGH
 DATA TOUR,TEXT,PICT,HCPY,ASSG,SPLT
 DATA 0,4,0,3,0,3,0,4,3,5,4,6,5,7,6,6
 RANDOMIZE(VAL(RIGHT$(TIME$,2)))
 SCREEN 1:CLS
 COLOR 0,0
 COLOR 1,1
 GOSUB MARKER
 GOSUB UNIFORM
 GOSUB MENUE
M1:
 GOSUB SELECTMENUE
 ON MC GOSUB SGINP,SNINP,SIO,SDEF,SRUN,SSTOP,SLOW,SHIGH,STOUR,
             STEXT,SPICT,SHCPY,DRAWASSIGNMENTS,SPLIT
 GOTO M1
END
```

```
REM --------------------------- CHANGE DEFAULT VALUES --------
SDEF:
 SCREEN 1:CLS
 NC%=NC%+1
 PRINT "# PARTITIONS        =";NC%;:INPUT VA:IF VA<>0 THEN NC%=VA
 PRINT "SELECTION PRESSURE =";RC;:INPUT VA:IF VA<>0 THEN RC=VA
 PRINT "MUTATION RATE      =";FR;:INPUT VA:IF VA<>0 THEN FR=VA
 PRINT "THRESHOLD (0)      =";TH;:INPUT VA:IF VA<>0 THEN TH=VA
 PRINT "SEL. PRESS. CHANGE =";DC;:INPUT VA:IF VA<>0 THEN DC=VA
 PRINT "DEFAULT DEVICE     =";" "+DEV$;:INPUT R$:IF R$<>"" THEN DEV$=R$
 NC%=NC%-1
 SCREEN 1:CLS
 GOSUB MENUE
RETURN
```

```
REM --------------------------- MENUE ------------------------
MENUE:
 FOR I%=1 TO 14
     LOCATE I%,35
     PRINT M$(I%)
 NEXT I%
RETURN
```

```
REM ------------------------- COMPUTE ASSIGNMENTS ------------
SRUN:
 GOSUB NORMALIZATION
RUN2:
 TRIALS%=TRIALS%+1
 LOCATE 20,30
 PRINT TRIALS%;
 FOR K%=0 TO NC%
     S=0
     FOR I%=0 TO N%
         IF X(I%,K%)<>0 THEN S=S+X(I%,K%)
     NEXT I%
     XH(K%)=S
 NEXT K%
 FOR I%=0 TO N%
     FOR K%=0 TO NC%
         S=0
         FOR J%=0 TO N%
             IF X(J%,K%)<>0 THEN S=S+G(I%,J%)*X(J%,K%)
         NEXT J%
         F(I%,K%)=S
     NEXT K%
 NEXT I%
 IF RC<0 GOTO RUN10
 FOR K%=0 TO NC%
     L%=K%-1
     R%=K%+1
     IF L%<0 THEN L%=L%+(NC%+1)
     IF R%>NC% THEN R%=R%-(NC%+1)
     FOR I%=0 TO N%
         IF X(I%,K%)<>0 THEN X(I%,K%)=X(I%,K%)*
         (1/(F(I%,L%)+F(I%,K%)+F(I%,R%))+RC*X(I%,K%))/XH(K%)
     NEXT I%
 NEXT K%
 GOTO RUN11
RUN10:
 FOR K%=0 TO NC%
     L%=K%-1
     R%=K%+1
     IF L%<0 THEN L%=L%+(NC%+1)
     IF R%>NC% THEN R%=R%-(NC%+1)
     FOR I%=0 TO N%
         IF X(I%,K%)<>0 THEN X(I%,K%)=X(I%,K%)*
         (1/(F(I%,L%)+F(I%,K%)+F(I%,R%))-RC*(XH(K%)-X(I%,K%)))/XH(K%)
     NEXT I%
 NEXT K%
RUN11:
 GOSUB NORMALIZATION
 PHIS=phi
 GOSUB MUTATIONS
 GOSUB DRAWASSIGNMENTS
 LOCATE 22,20 PRINT PHIS;
 R$=INKEY$:IF R$<>"" THEN RETURN
 GOTO RUN2
RETURN
```

```
REM ------------------------- COMPUTE TOUR --------------------
STOUR:
 IF N%=NC% GOTO TOUR4
 SCREEN 1:CLS
 GOSUB DRAWCITIES
 GOSUB MENUE
RETURN
TOUR4:
 SWITCH%=1
 FOR K%=0 TO N%
     S=0
     FOR I%=0 TO N%
          IF X(I%,K%)<=S GOTO TOUR1
          S=X(I%,K%)
          TA(K%)=I%
TOUR1:
     NEXT I%
     S=0
     FOR L%=0 TO N%
          IF X(TA(K%),L%)<=S GOTO TOUR3
          S=X(TA(K%),L%)
          HA%=L%
TOUR3:
     NEXT L%
     IF HA%=K% GOTO TOUR2
     TA(K%)=-1
     SWITCH%=0
TOUR2:
 NEXT K%
 SCREEN 1:CLS
 GOSUB MENUE
 GOSUB DRAWSTREETS
 GOSUB DRAWCITIES
REM -------------- COMPUTE TOUR LENGTH ------------------------
 IF SWITCH%=0 THEN RETURN
 TL=0
 FOR I%=0 TO N%-1
     TL=TL+G(TA(I%),TA(I%+1))
 NEXT I%
 TL=TL+G(TA(N%),TA(0))
 LOCATE 1,1
 PRINT "TOUR LENGTH ="+STR$(TL)
RETURN
```

```
REM --------------------- UNIFORM DISTRIBUTION ----------------
UNIFORM:
 FOR I%=0 TO N%
     FOR K%=0 TO NC%
          X(I%,K%)=1
     NEXT K%
 NEXT I%
RETURN
```

```
REM ------------------- MUTATIONS -----------------------------
MUTATIONS:
 FOR I%=0 TO N%
     FOR K%=0 TO NC%
          IF X(I%,K%)<>0 THEN X(I%,K%)=X(I%,K%)+RND(1)*FR
     NEXT K%
 NEXT I%
 GOSUB NORMALIZATION
RETURN
```

```
REM --------------- NORMALIZATION ----------------------------
NORMALIZATION:
 phi=0
 FOR I%=0 TO N%
     S=0
     FOR K%=0 TO NC%
         IF X(I%,K%)<>0 THEN S=S+X(I%,K%)
     NEXT K%
     phi=phi+S
     FOR K%=0 TO NC%
         IF X(I%,K%)<>0 THEN X(I%,K%)=X(I%,K%)/S
     NEXT K%
 NEXT I%
RETURN
```

```
REM -------------------- SELECT MENUE ------------------------
SELECTMENUE:
 PUT (MX,MY),MARK,XOR
 C$=CHR$(56)+CHR$(50)+CHR$(13)
MARKER1:
 R$=INKEY$:IF R$="" GOTO MARKER1
 H=INSTR(C$,R$)
 IF H=0 GOTO MARKER1
 ON H GOTO MARKER2,MARKER3,MARKER4
MARKER2:
 IF MY<=3 GOTO MARKER1
 PUT (MX,MY),MARK,XOR
 MY=MY-8
 GOTO MARKER5
MARKER3:
 IF MY>=107
 GOTO MARKER1
 PUT (MX,MY),MARK,XOR
 MY=MY+8
MARKER5:
 PUT (MX,MY),MARK,XOR
 GOTO MARKER1
MARKER4:
 PUT (MX,MY),MARK,XOR
 MC=(MY-3)/8+1
RETURN
```

```
REM ---------------- DRAW ASSIGNMENTS ------------------------
DRAWASSIGNMENTS:
 SCREEN 1:CLS
 GOSUB MENUE
 FOR I%=0 TO N%
     FOR K%=0 TO NC%
         S=INT(X(I%,K%)*4)
         X1=20+K%*4-INT(X(I%,K%)*2)
         Y1= 2+I%*4-INT(X(I%,K%)*2)
         LINE (X1,Y1)-(X1+S,Y1+S),3,BF
     NEXT K%
 NEXT I%
 LOCATE 20,30
 PRINT TRIALS%;
RETURN
```

```
REM ------------------------- SPLIT PARTITIONS ----------------
SPLIT:
 FOR I%=0 TO N%
      FOR K%=0 TO NC%
           IF X(I%,K%)<=TH THEN X(I%,K%)=0
      NEXT K%
 NEXT I%
 GOSUB NORMALIZATION
 TAMAX=0
 TAMIN=10000
 FOR K%=0 TO NC%
      TA(K%)=0
      FOR I%=0 TO N%
           TA(K%)=TA(K%)+X(I%,K%)
      NEXT I%
      TA(K%)=INT(TA(K%)+.5)
      IF TA(K%)>TAMAX THEN TAMAX=TA(K%)
      IF TA(K%)<TAMIN THEN TAMIN=TA(K%)
 NEXT K%
 SCREEN 1:CLS
 PRINT "SPLIT A PARTITION INTO # PARTITIONS =";
 INPUT SP%
 IF SP%>TAMAX THEN SP%=TAMAX
 NCN%=0
 FOR K%=0 TO NC%
      IF SP%> TA(K%) THEN NCN%=NCN%+TA(K%)
      IF SP%<=TA(K%) THEN NCN%=NCN%+SP%
 NEXT K%
 NCN%=NCN%-1
 IF NCN%<=N% GOTO SPLIT1
 PRINT "# CITIES TOO LARGE !"
 PRINT NCN%+1
SPLIT10:
 R$=INKEY$:IF R$="" GOTO SPLIT10
RETURN
SPLIT1:
 LO%=NCN%
 LU%=LO%+1
 FOR K%=NC% TO 0 STEP -1
      LO%=LU%-1
      IF SP%<=TA(K%) THEN LU%=LO%-(SP%-1)
      IF SP%> TA(K%) THEN LU%=LO%-(TA(K%)-1)
      IF LO%=LU% GOTO SPLIT2
      FOR L%=LO% TO LU% STEP -1
           FOR I%=0 TO N%
                X(I%,L%)=X(I%,K%)
           NEXT I%
      NEXT L%
      GOTO SPLIT3
SPLIT2:
      FOR I%=0 TO N%
           X(I%,LO%)=X(I%,K%)
      NEXT I%
SPLIT3:
 NEXT K%
 NC%=NCN%
 GOSUB NORMALIZATION
 SCREEN 1:CLS
 GOSUB MENUE
RETURN
```

Index

H

I

J

L

M

N

P

R

S

Bei Fragen zur Produktsicherheit wenden Sie sich bitte an:
If you have any questions regarding product safety,
please contact:

Walter de Gruyter GmbH
Genthiner Straße 13
10785 Berlin
productsafety@degruyterbrill.com